TAKE NOTE!

TO ACCOMPANY

CELL AND MOLECULAR BIOLOGY
CONCEPTS AND EXPERIMENTS

Third Edition

Gerald Karp

JOHN WILEY & SONS, INC.

To order books or for customer service call 1-800-CALL-WILEY (225-5945).

Copyright © 2002 John Wiley & Sons, Inc. All rights reserved.

No part of this publication may be reproduced, stored in a retrieval system or transmitted in any form or by any means, electronic, mechanical, photocopying, recording, scanning or otherwise, except as permitted under Sections 107 or 108 of the 1976 United States Copyright Act, without either the prior written permission of the Publisher, or authorization through payment of the appropriate per-copy fee to the Copyright Clearance Center, 222 Rosewood Drive, Danvers, MA 01923, (978) 750-8400, fax (978) 750-4470. Requests to the Publisher for permission should be addressed to the Permissions Department, John Wiley & Sons, Inc., 605 Third Avenue, New York, NY 10158-0012, (212) 850-6011, fax (212) 850-6008, E-Mail: PERMREQ@WILEY.COM.

ISBN 0-471-12775-2

Printed in the United States of America

10 9 8 7 6 5 4 3 2 1

Printed and bound by Courier Westford, Inc.

CONTENTS

1	Figure 1.9 b,c	22	Figure 5.26		Figure 9.26		Figure 12.14	79	Figure 15.28
2	Figure 1.21		Figure 5.29	41	Figure 9.31b	61	Figure 12.16a		Figure 15.30
	Figure EP.1	23	Figure 5.30		Figure 9.37		Figure 12.19c	80	Figure 15.36
3	Figure 2.12		Figure 6.4		Figure 9.38a	62	Figure 12.26		Figure 15.37
	Figure 2.19a	24	Figure 6.9	42	Figure 9.57a	63	Figure 12.29	81	Figure 16.11
4	Figure 2.26		Figure 6.10a		Figure 9.58a		Figure 12.30a		Figure 16.13
5	Figure 2.30	25	Figure 6.14	43	Figure 9.62	64	Figure 12.32	82	Figure 17.2
	Figure 2.31		Figure 6.15a		Figure 9.63		Figure 12.34		Figure 17.5
6	Figure 2.34a	26	Figure 6.18	44	Figure 9.66	65	Figure 12.47	83	Figure 17.9
	Figure 2.48	27	Figure 6.19		Figure 9.73		Figure 12.59		Figure 17.25
7	Figure 3.8	28	Figure 7.1	45	Figure 10.10a,b	66	Figure 13.3a	84	Figure 17.15
	Figure 3.9		Figure 7.5	46	Figure 10.14b		Figure 13.4a		Figure 17.16
8	Figure 3.21	29	Figure 7.17c		Figure 10.26	67	Figure 13.8	85	Figure 18.16
	Figure 3.22		Figure 7.18b	47	Figure 10.30		Figure 13.11		Figure 18.20
9	Figure 3.23		Figure 7.26	48	Figure EP. 2	68	Figure 13.13	86	Figure 18.23
	Figure 3.28	30	Figure 7.27b		Figure EP. 4		Figure 13.22		Figure 18.27
10	Figure 4.4		Figure 7.30b	49	Figure 11.1	69	Figure 13.27	87	Figure 18.34
11	Figure 4.6		Figure 7.32b		Figure 11.2		Figure 13.28		Figure 18.37
	Figure 4.12	31	Figure 8.2	50	Figure 11.4a-c	70	Figure 14.1	88	Figure 18.38
12	Figure 4.17		Figure 8.4c	51	Figure 11.13		Figure 14.6		Figure 18.45
	Figure 4.22	32	Figure 8.5a		Figure 11.17		Figure 14.9	89	Figure 18.47
13	Figure 4.29		Figure 8.7a	52	Figure 11.19	71	Figure 14.11		Figure 18.48
	Figure 4.30	33	Figure 8.11a		Figure 11.21	72	Figure 14.23		
14	Figure 4.32d		Figure 8.14	53	Figure 11.28		Figure 14.25		
	Figure 4.33	34	Figure 8.12		Figure 11.29	73	Figure 14.30		
15	Figure 4.38		Figure 8.13	54	Figure 11.32b		Figure 14.38		
	Figure 4.41	35	Figure 8.17		Figure 11.39	74	Figure 14.40		
16	Figure 4.44	36	Figure 8.25	55	Figure 11.35		Figure 14.47		
	Figure 4.47		Figure 8.28	56	Figure 11.41	75	Figure 15.3b		
17	Figure 4.50a	37	Figure 8.29b	57	Figure 11.42a		Figure 15.7		
	Figure 4.54		Figure 8.39a		Figure 11.47	76	Figure 15.10		
18	Figure 5.5	38	Figure 8.47a	58	Figure 11.49		Figure 15.21		
	Figure 5.16		Figure 8.48		Figure 11.50a	77	Figure 15.12		
19	Figure 5.6	39	Figure 9.1	59	Figure 12.2a		Figure 15.15		
20	Figure 5.7		Figure 9.16 a,b		Figure 12.8a	78	Figure 15.23		
21	Figure 5.11	40	Figure 9.18 b,c	60	Figure 12.12b		Figure 15.25		

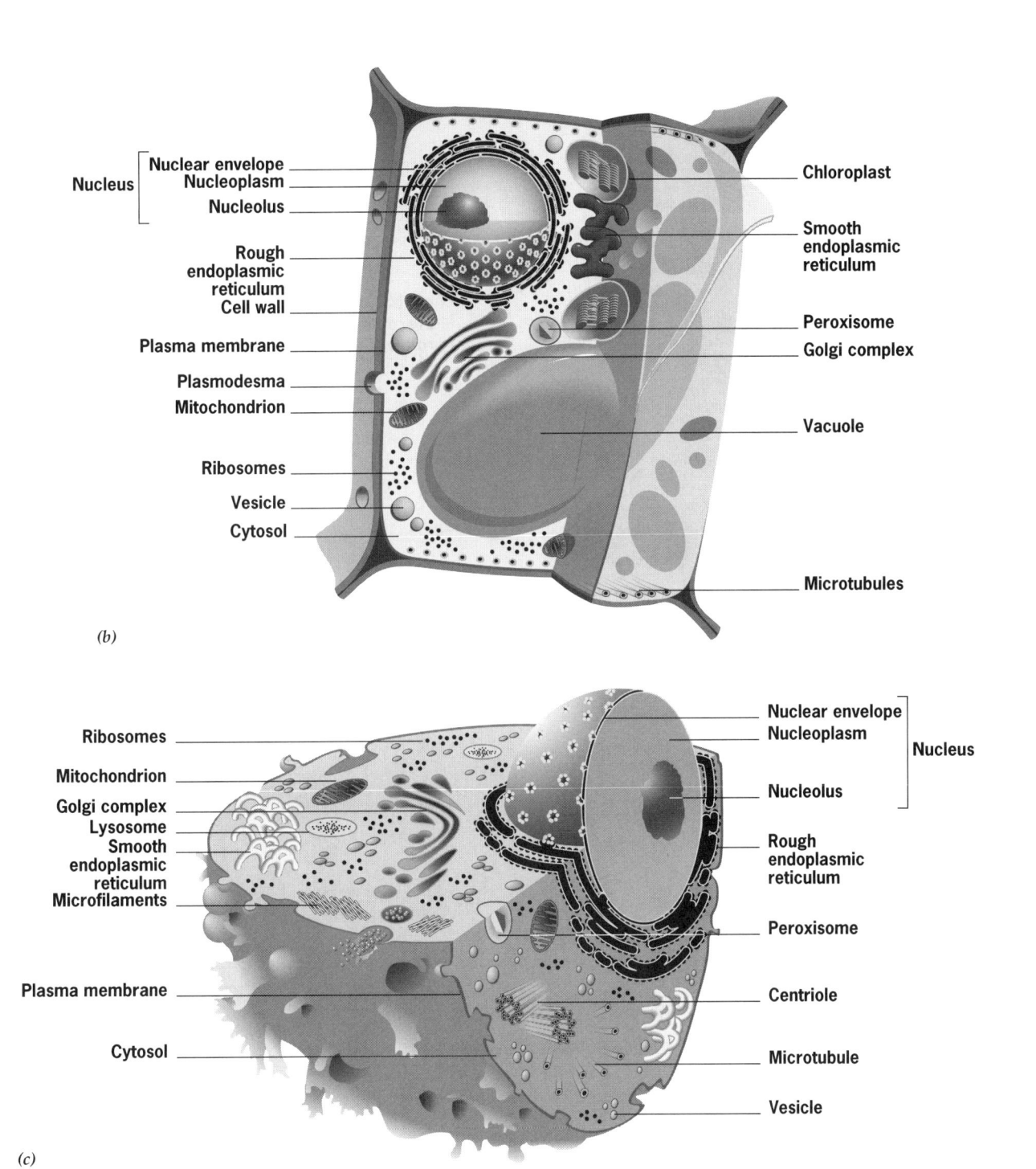

Figure 1.9b,c The structure of cells.

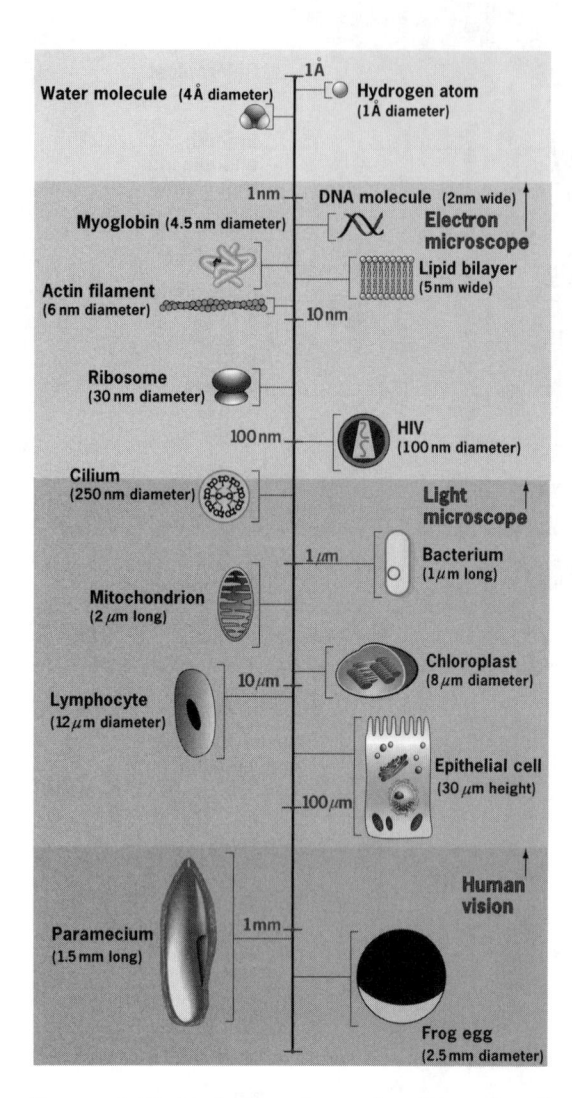

Figure 1.21 Relative sizes of cells and cell components.

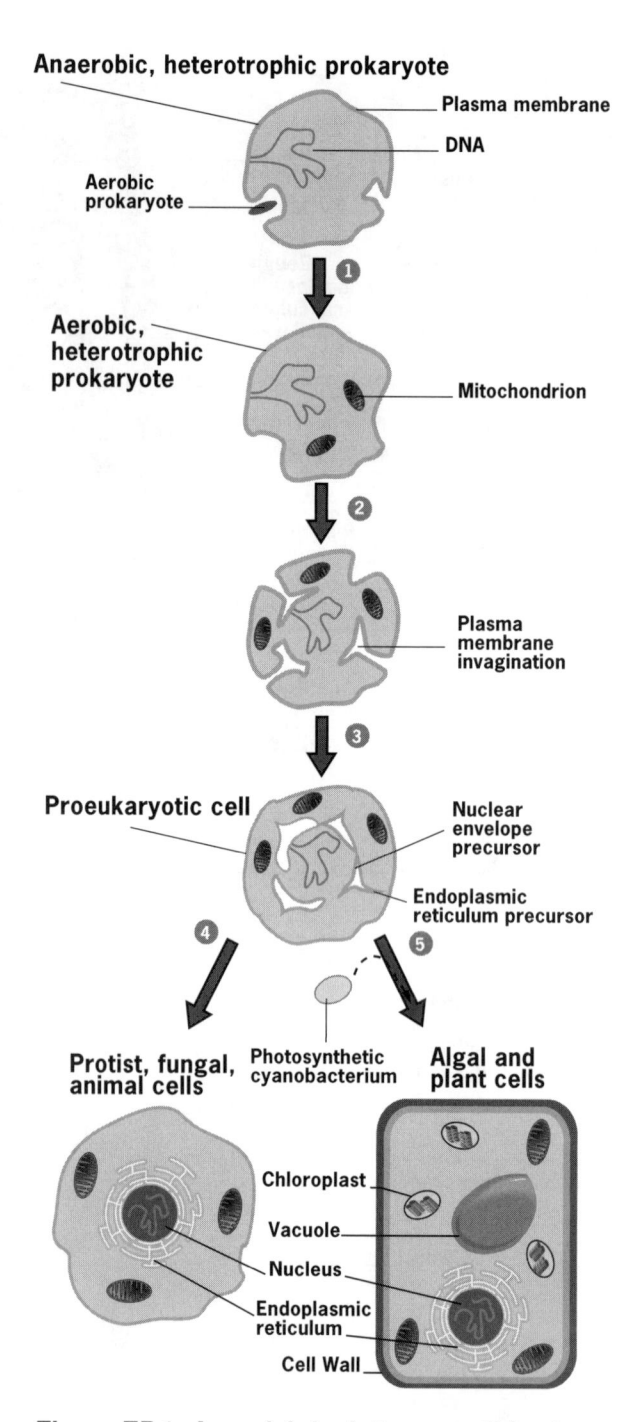

Figure EP.1 A model depicting possible steps in the evolution of eukaryotic cells, including the origin of mitochondria and chloroplasts by endosymbiosis.

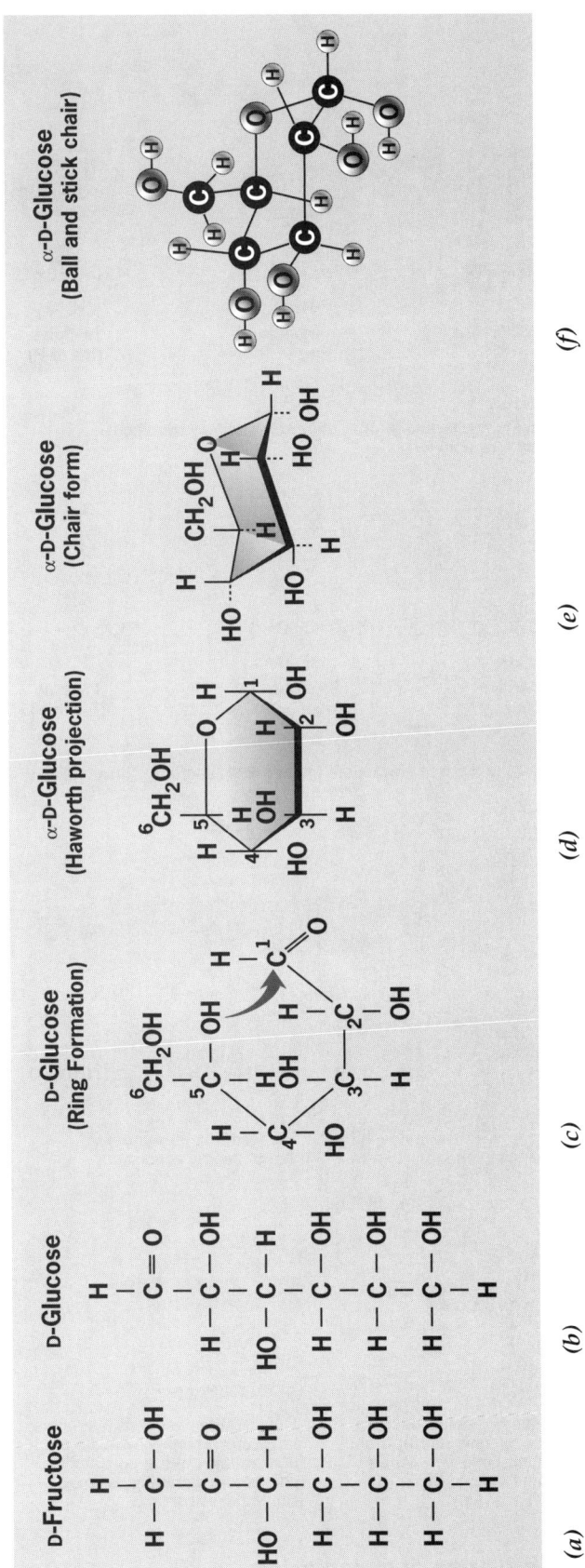

Figure 2.12 The structures of sugars.

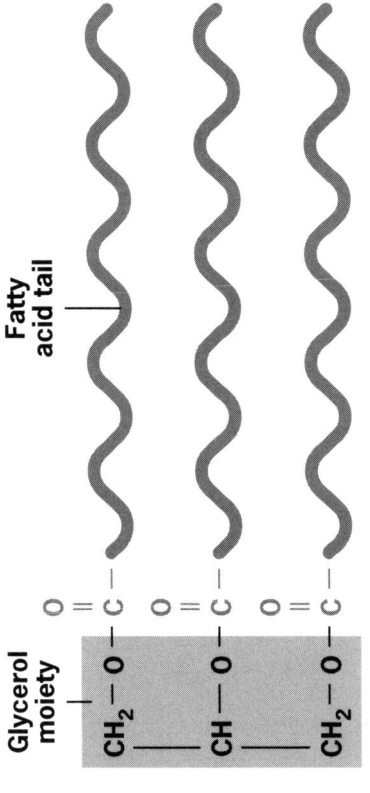

Figure 2.19a Fats and fatty acids.

Figures 2.12 & 2.19a

Polar charged

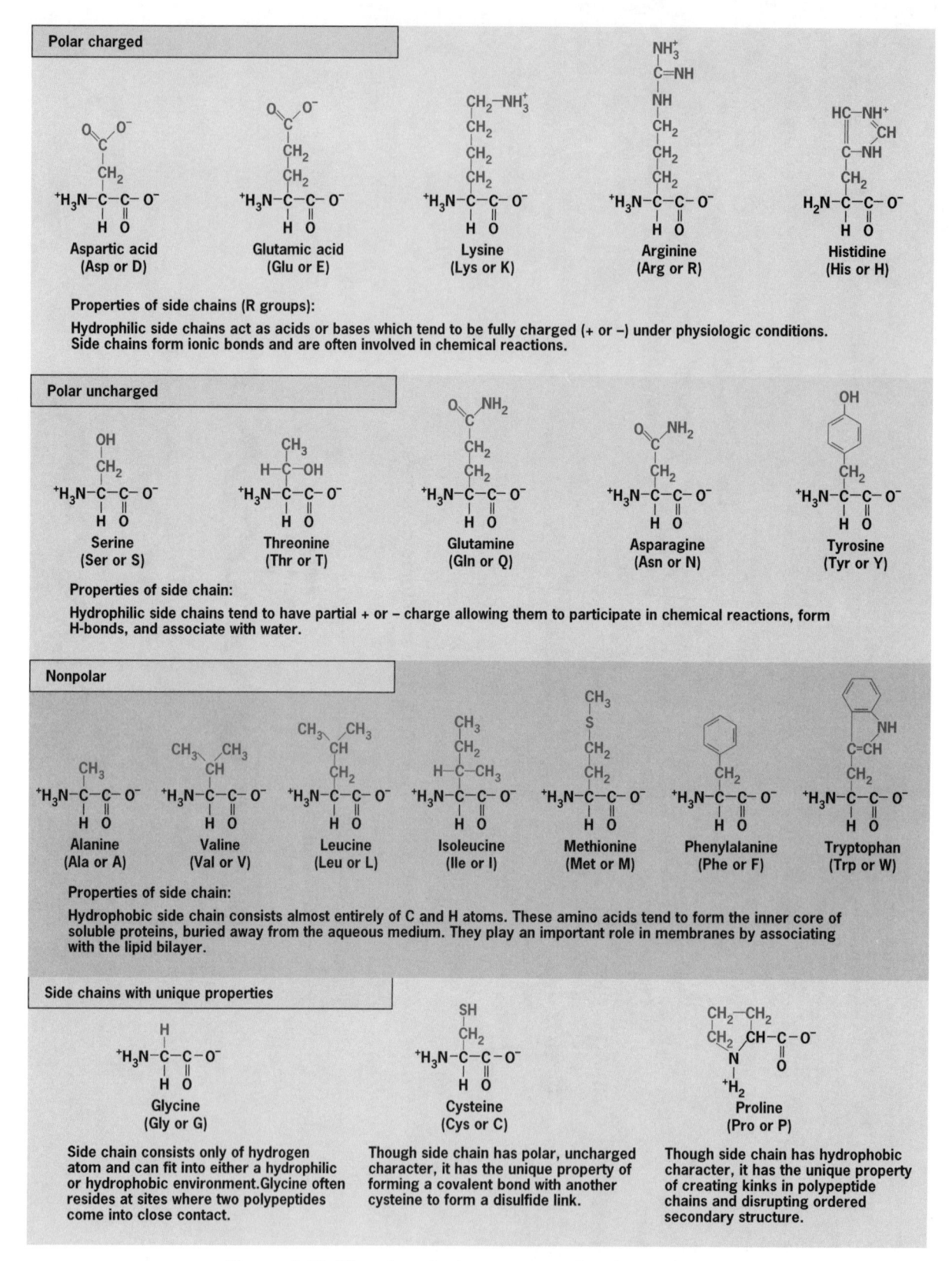

Aspartic acid (Asp or D) | Glutamic acid (Glu or E) | Lysine (Lys or K) | Arginine (Arg or R) | Histidine (His or H)

Properties of side chains (R groups):
Hydrophilic side chains act as acids or bases which tend to be fully charged (+ or −) under physiologic conditions. Side chains form ionic bonds and are often involved in chemical reactions.

Polar uncharged

Serine (Ser or S) | Threonine (Thr or T) | Glutamine (Gln or Q) | Asparagine (Asn or N) | Tyrosine (Tyr or Y)

Properties of side chain:
Hydrophilic side chains tend to have partial + or − charge allowing them to participate in chemical reactions, form H-bonds, and associate with water.

Nonpolar

Alanine (Ala or A) | Valine (Val or V) | Leucine (Leu or L) | Isoleucine (Ile or I) | Methionine (Met or M) | Phenylalanine (Phe or F) | Tryptophan (Trp or W)

Properties of side chain:
Hydrophobic side chain consists almost entirely of C and H atoms. These amino acids tend to form the inner core of soluble proteins, buried away from the aqueous medium. They play an important role in membranes by associating with the lipid bilayer.

Side chains with unique properties

Glycine (Gly or G) | Cysteine (Cys or C) | Proline (Pro or P)

Side chain consists only of hydrogen atom and can fit into either a hydrophilic or hydrophobic environment. Glycine often resides at sites where two polypeptides come into close contact.

Though side chain has polar, uncharged character, it has the unique property of forming a covalent bond with another cysteine to form a disulfide link.

Though side chain has hydrophobic character, it has the unique property of creating kinks in polypeptide chains and disrupting ordered secondary structure.

Figure 2.26 The chemical structure of amino acids.

Figure 2.31 The β-pleated sheet.

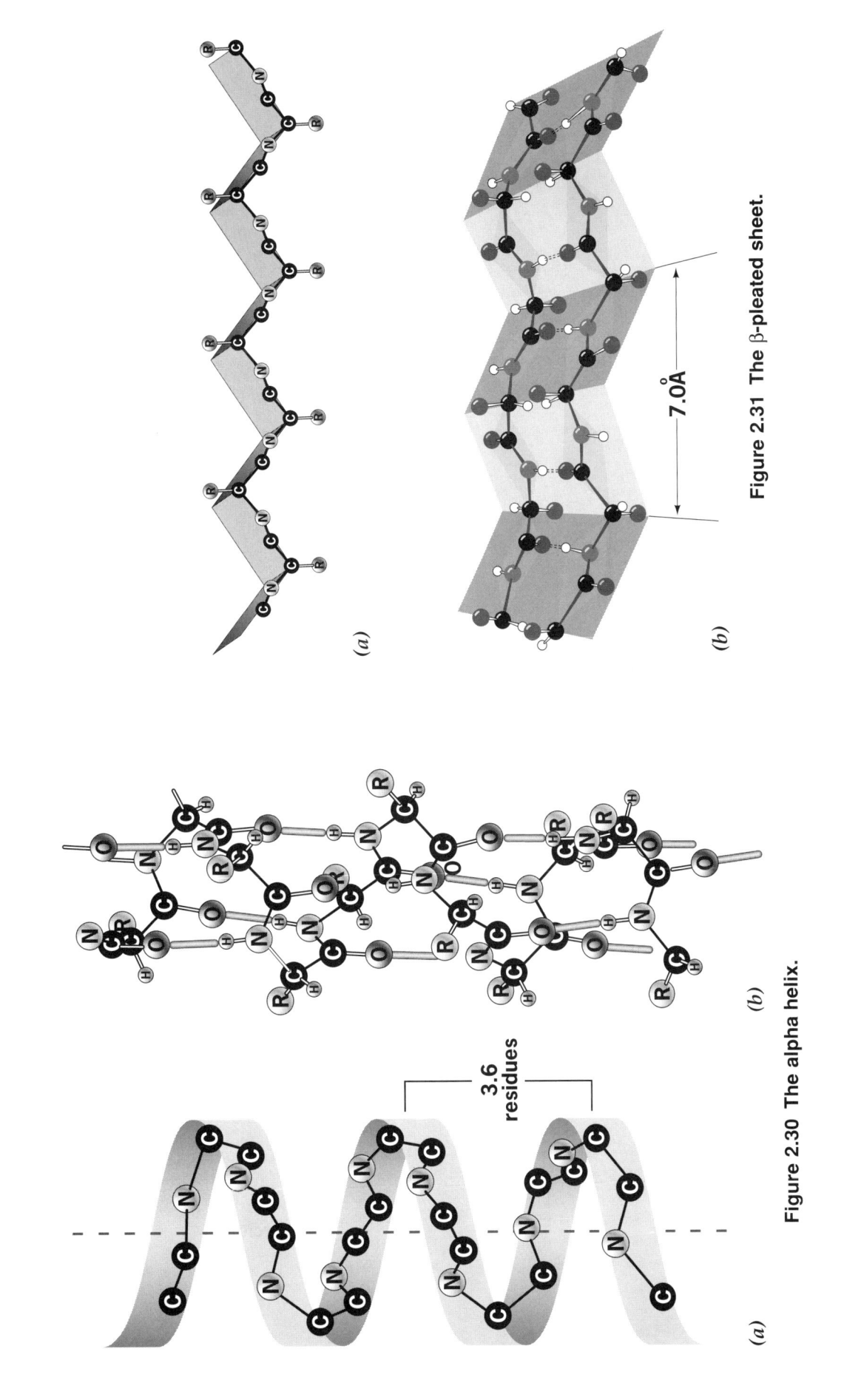

7.0Å

Figure 2.30 The alpha helix.

3.6 residues

Figures 2.30 & 2.31

(a)

Figure 2.34a The three-dimensional structure of whale myoglobin.

(b)

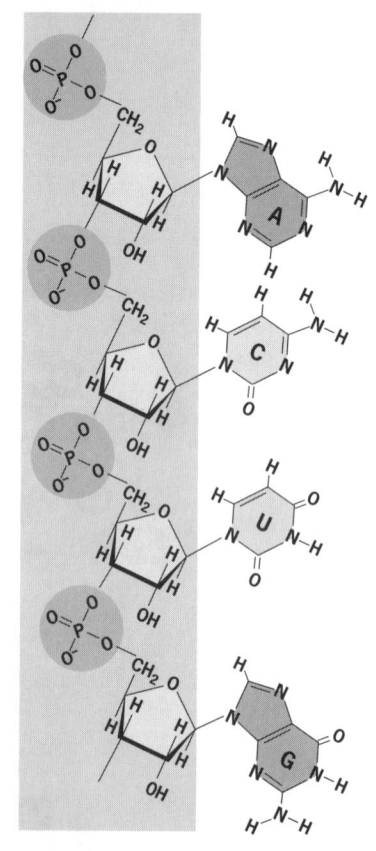

Figure 2.48 Nucleotides and nucleotide strands of RNA.

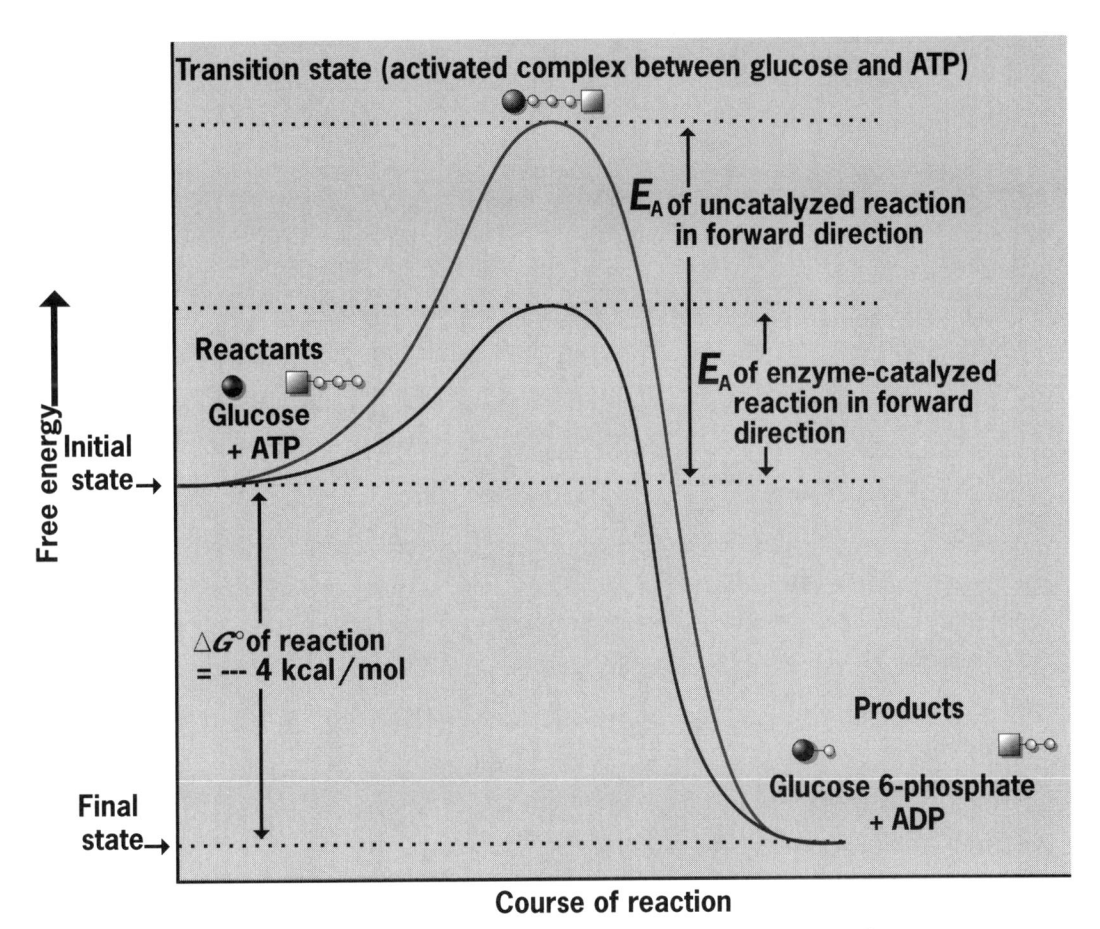

Figure 3.8 Activation energy and enzymatic reactions.

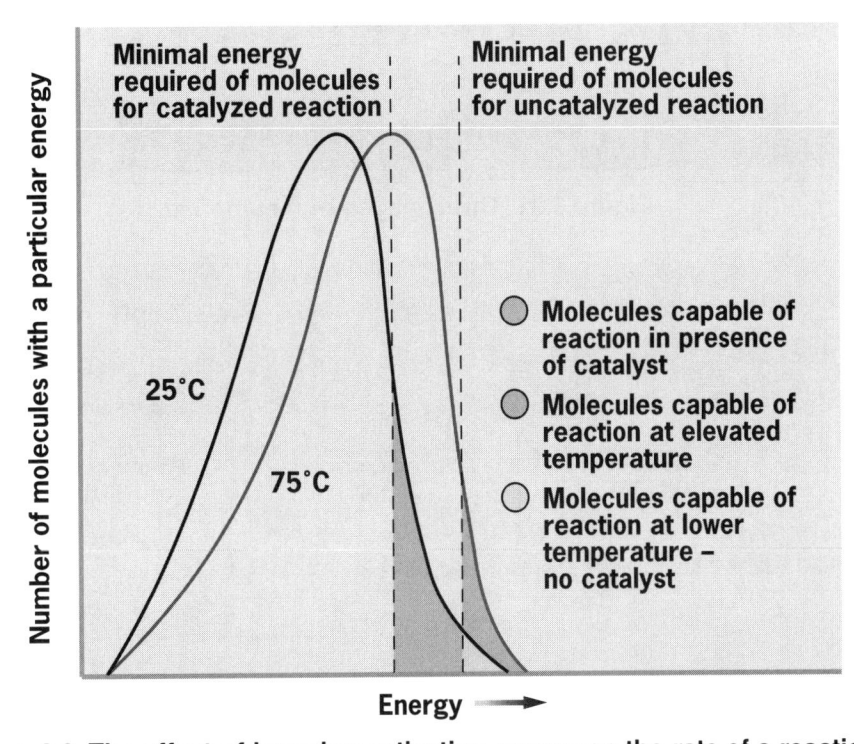

Figure 3.9 The effect of lowering activation energy on the rate of a reaction.

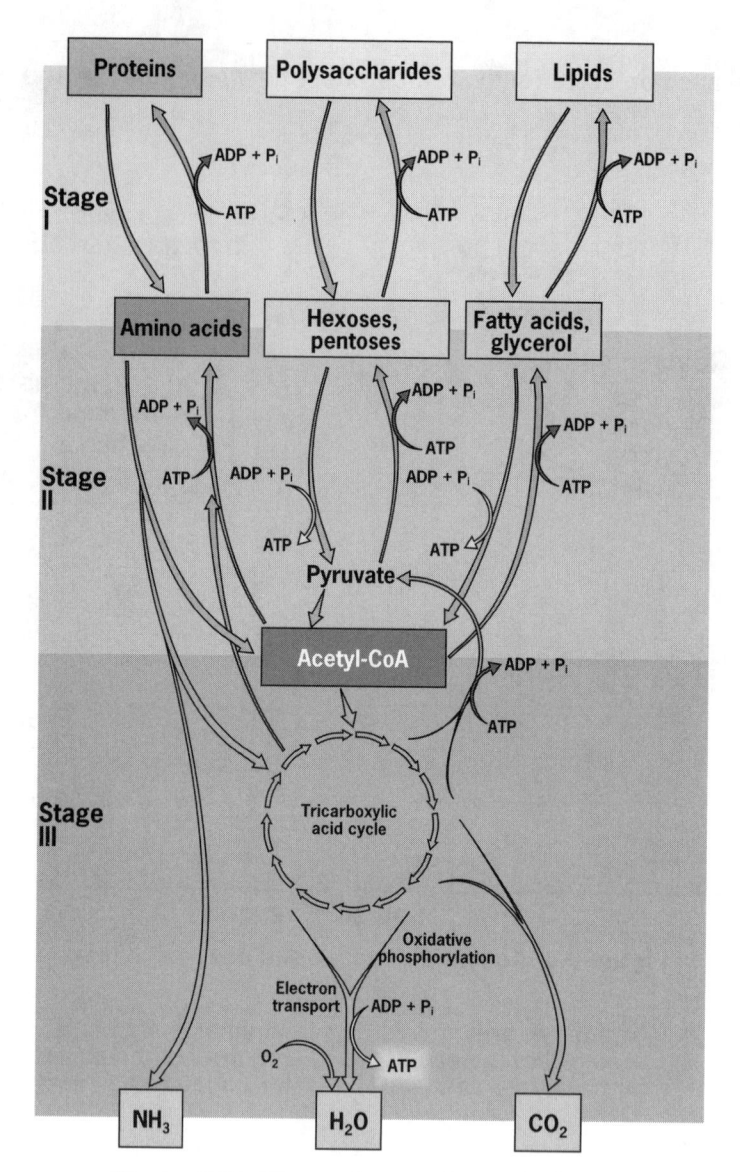

Figure 3.21 Three stages of metabolism.

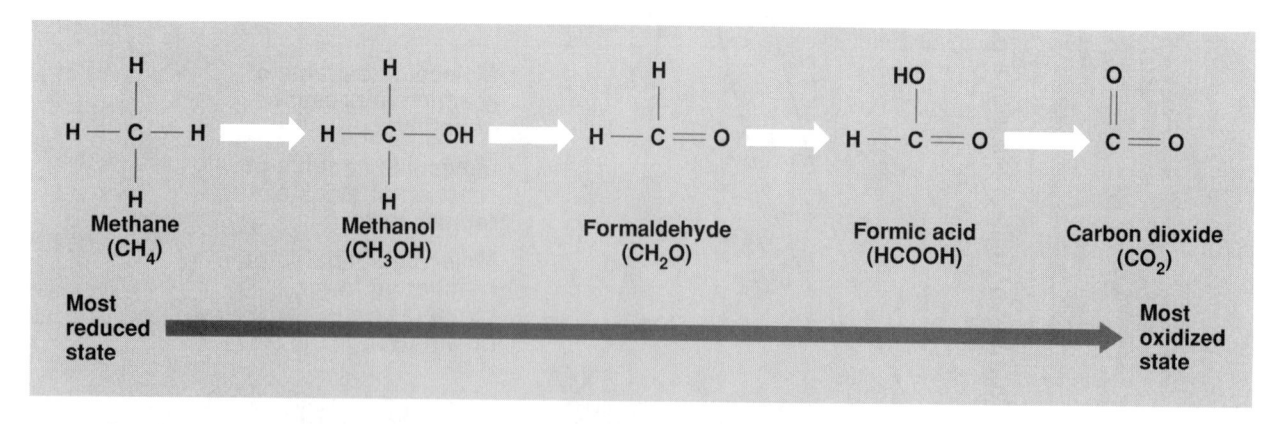

Covalent bond in which carbon atom has greater share of electron pair
Covalent bond in which oxygen atom has greater share of electron pair

Figure 3.22 The oxidation state of a carbon atom depends on the other atoms to which it is bonded.

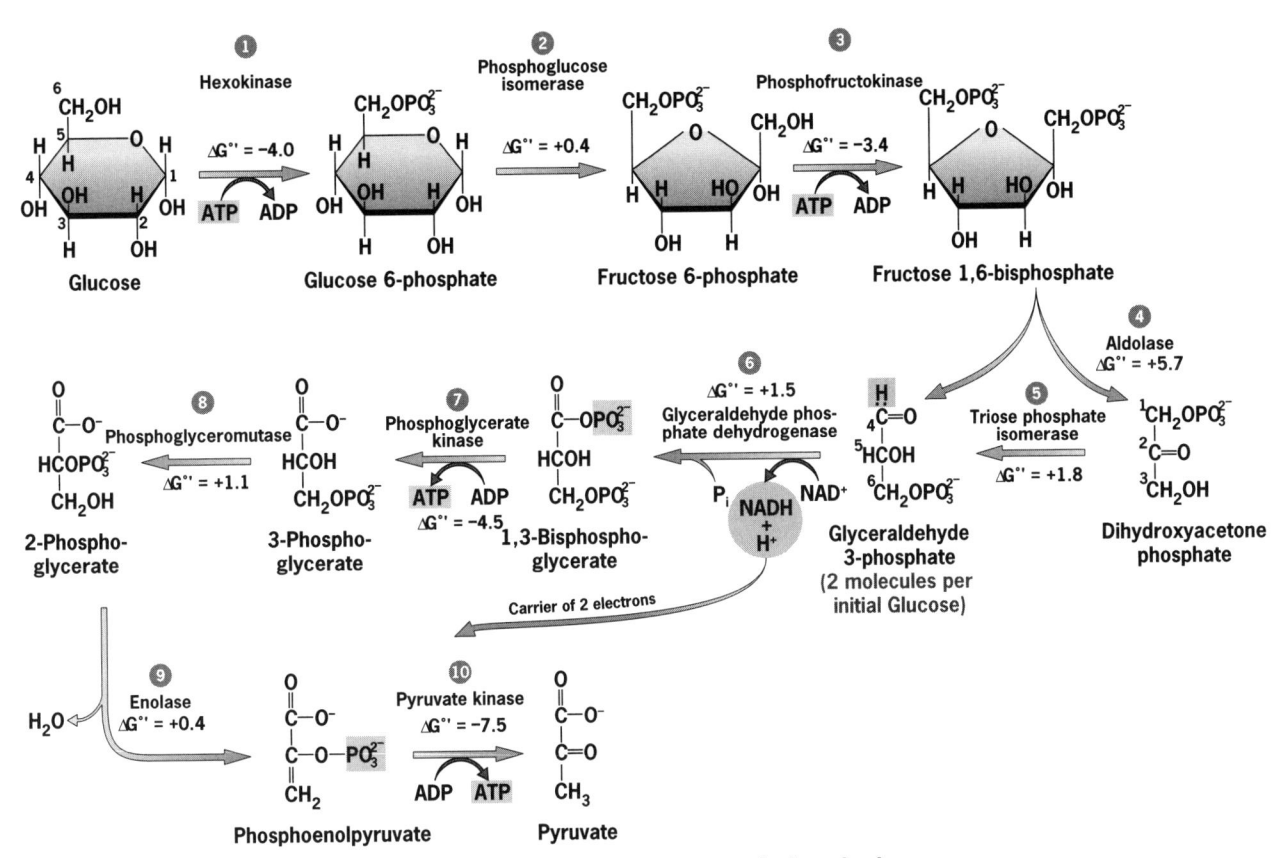

Figure 3.23 The steps of glycolysis.

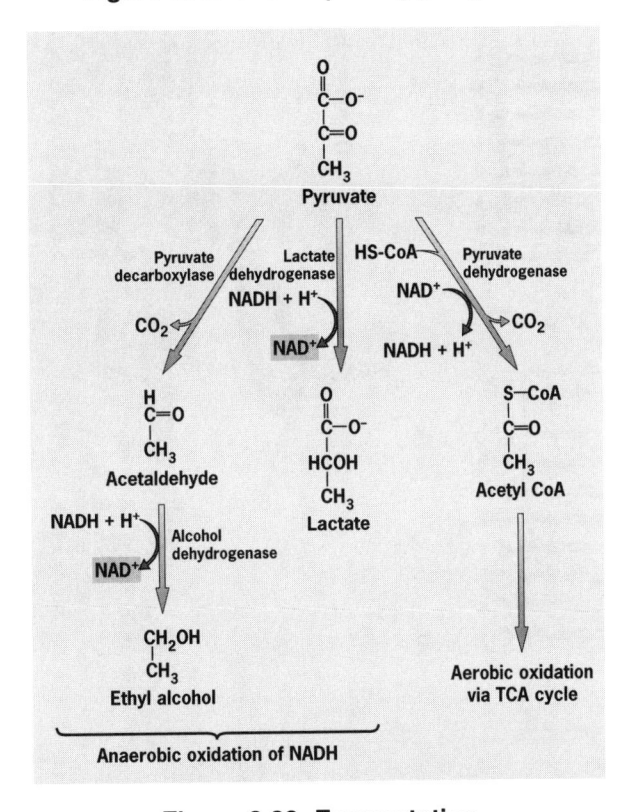

Figure 3.28 Fermentation.

Figures 3.23 & 3.28

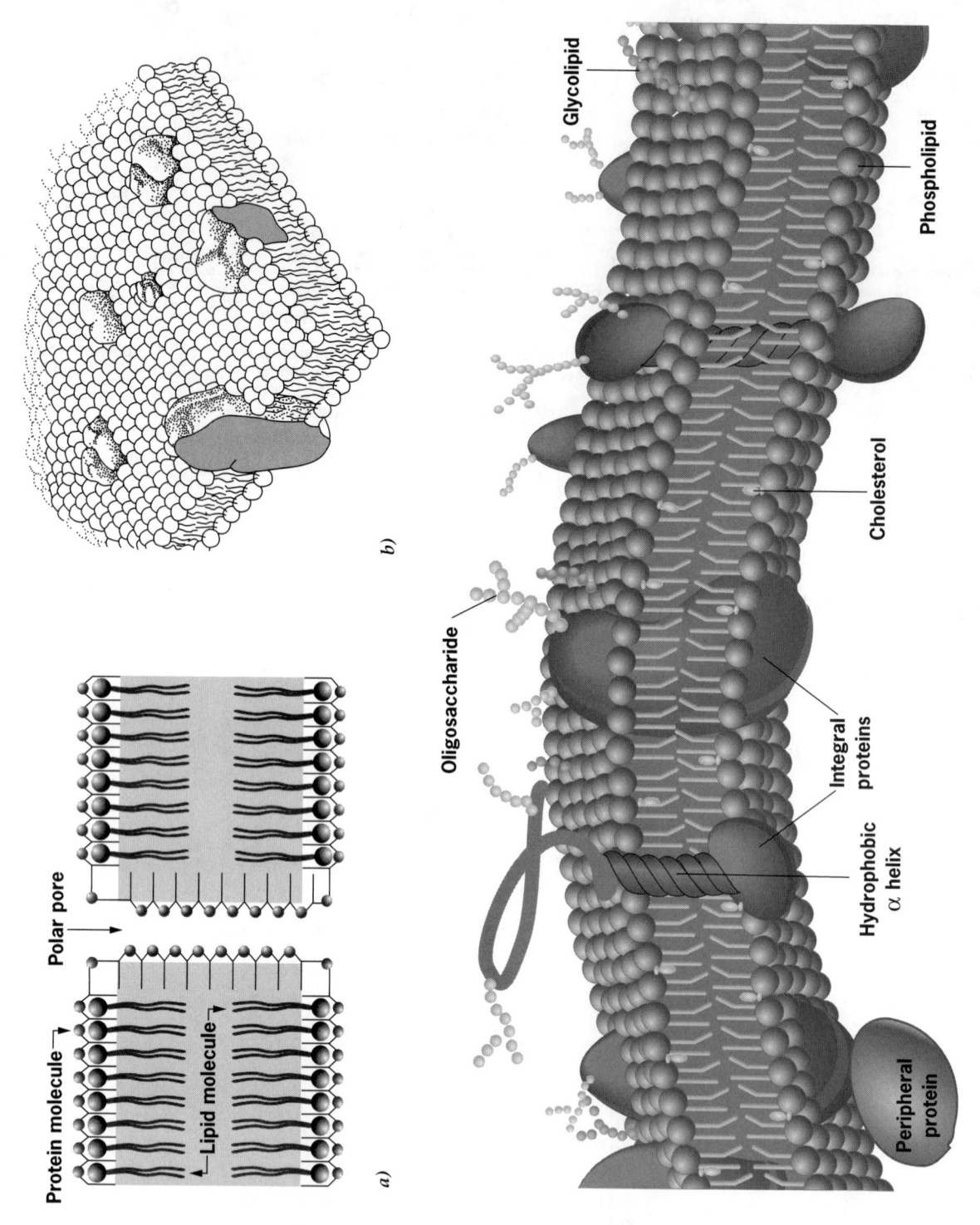

Figure 4.4 A brief history of the structure of the plasma membrane.

Figure 4.4

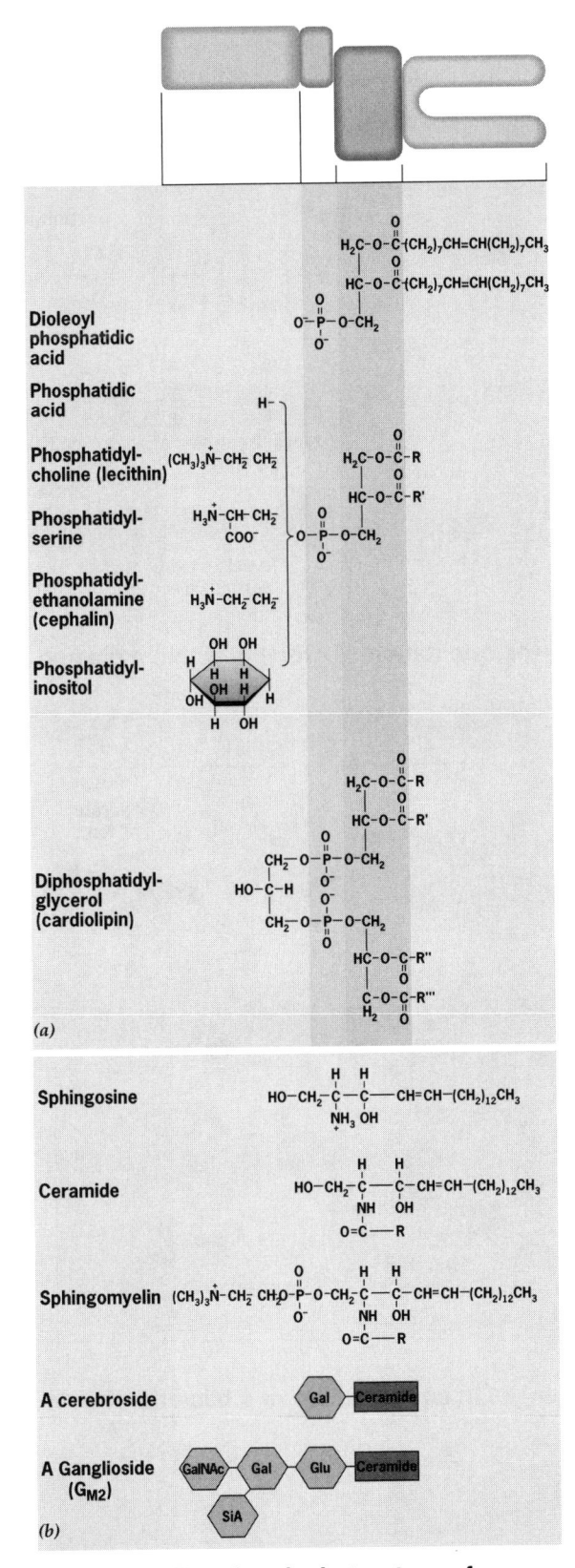

Figure 4.6 The chemical structure of membrane lipids.

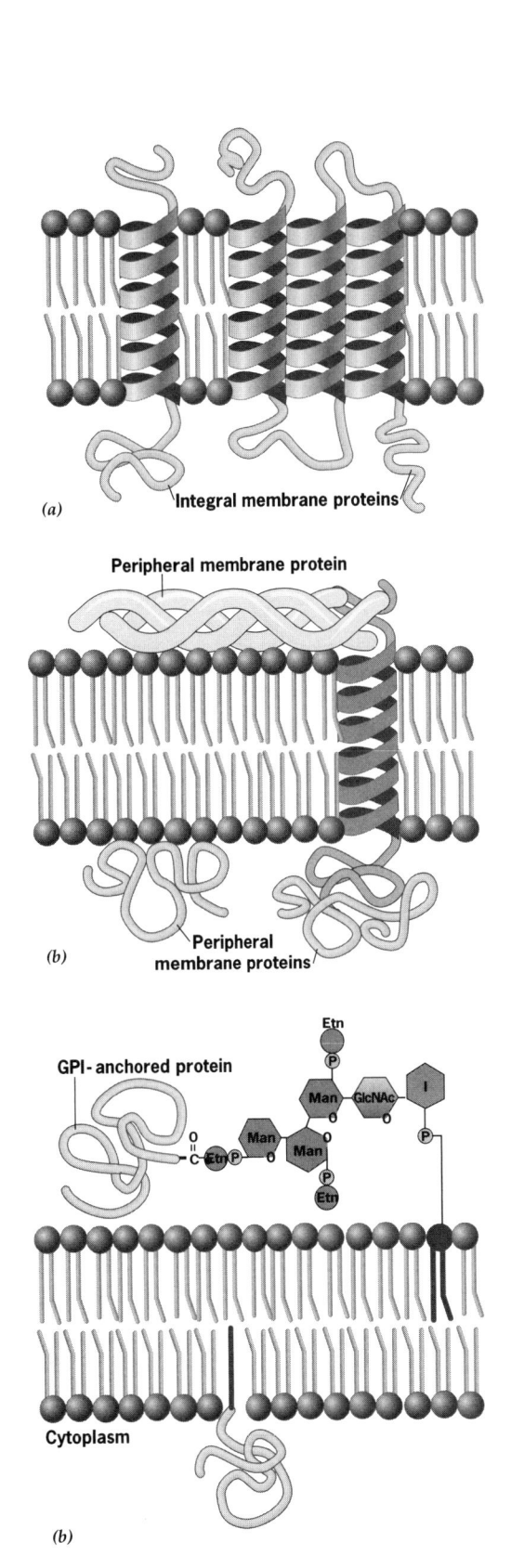

Figure 4.12 Three classes of membrane protein.

Figures 4.6 & 4.12

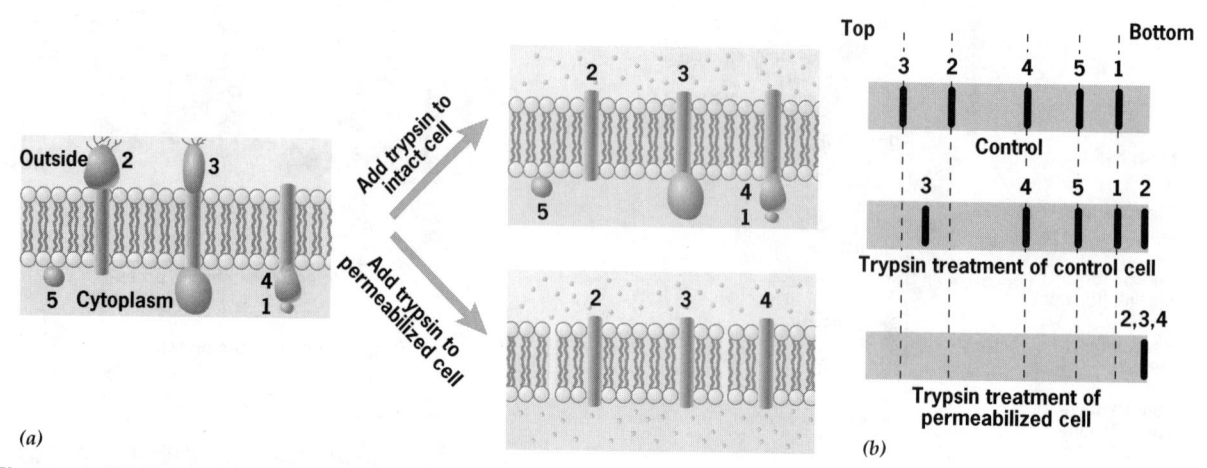

Figure 4.17 Experimental procedure for determining the orientation of proteins within a plasma membrane.

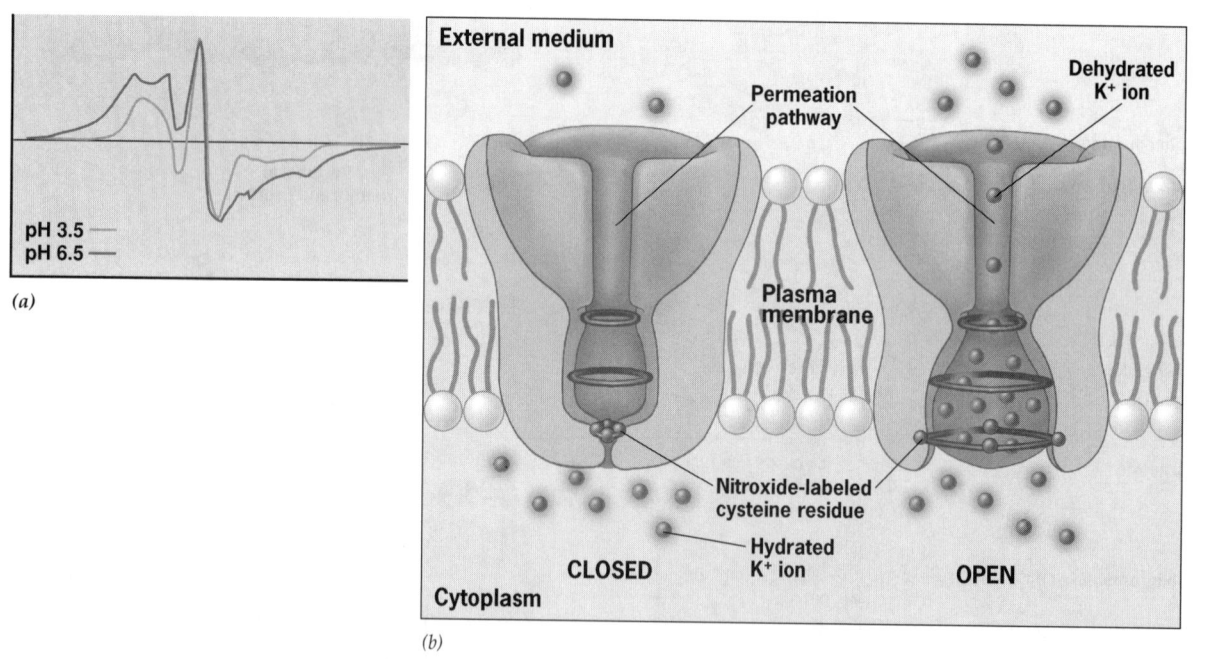

Figure 4.22 Use of EPR spectroscopy to monitor changes in conformation of a bacterial K⁺ ion channel as it opens and closes.

(a)

(b)

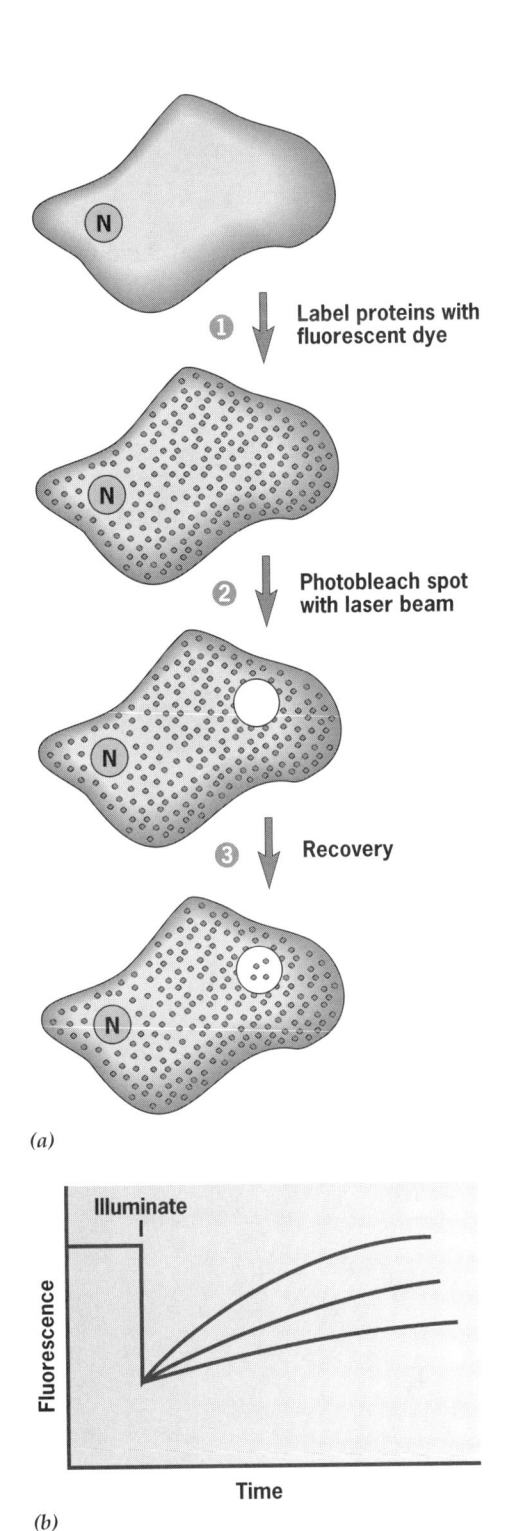

Figure 4.29 Measuring the diffusion rates of membrane proteins by fluorescence recovery after photobleaching (FRAP).

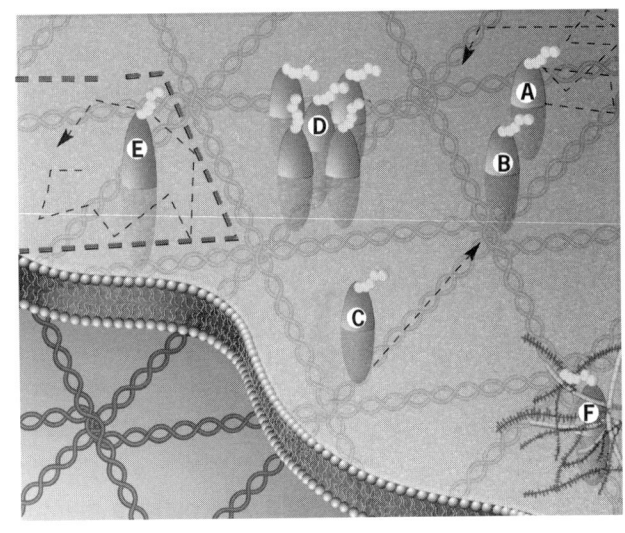

Figure 4.30 Patterns of movement of integral membrane proteins.

Figures 4.29 & 4.30

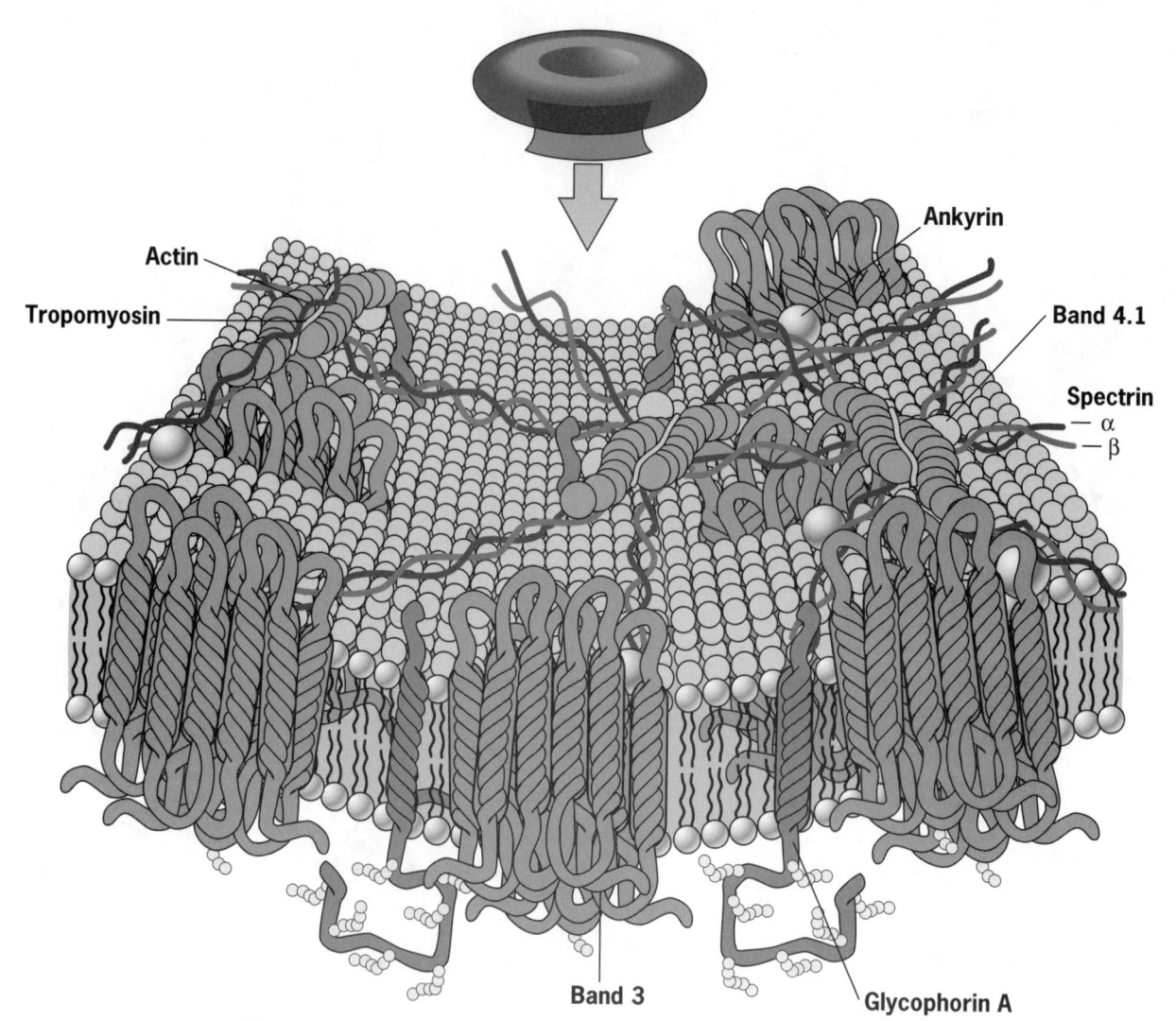

Figure 4.32d The plasma membrane of the human erythrocyte.

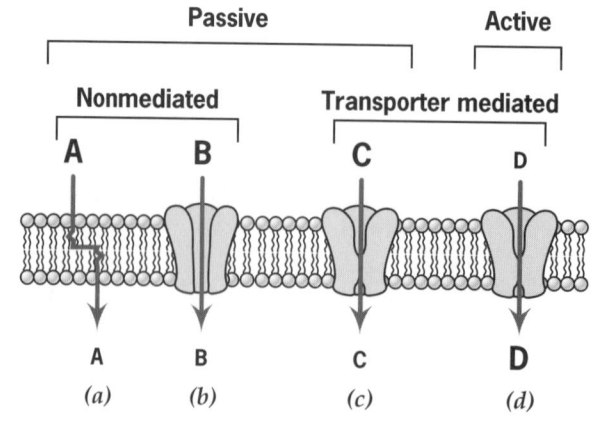

Figure 4.33 Four basic mechanisms by which solute molecules move across membranes.

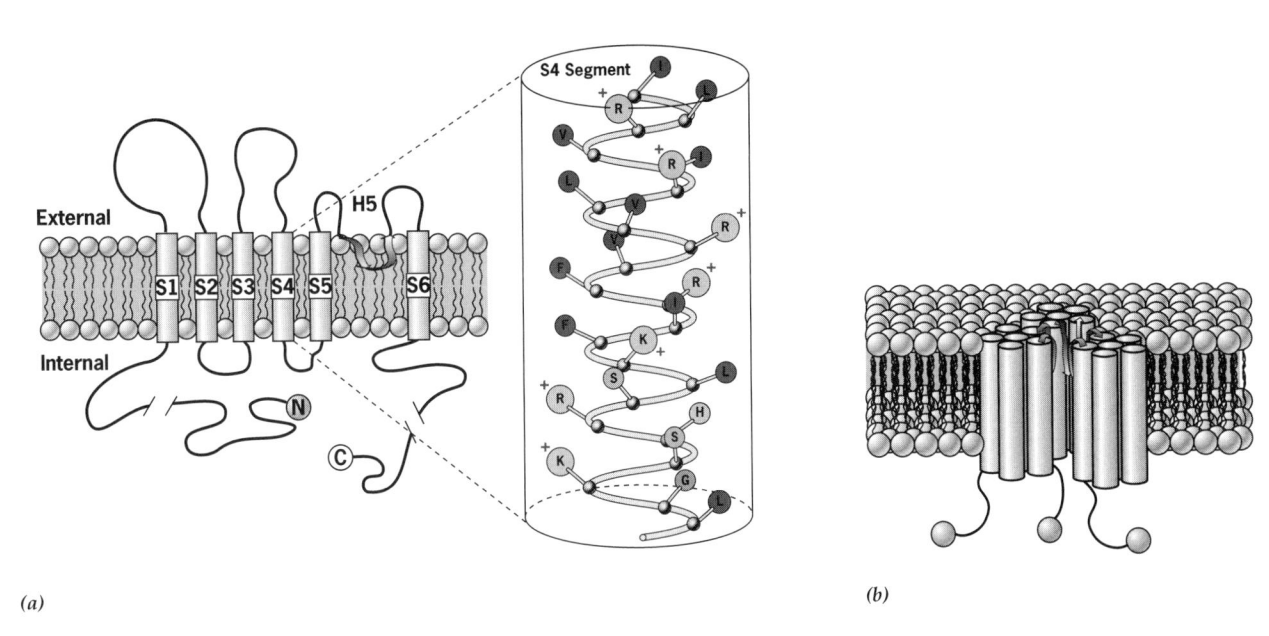

Figure 4.38 The structure of a voltage-gated K$^+$ channel.

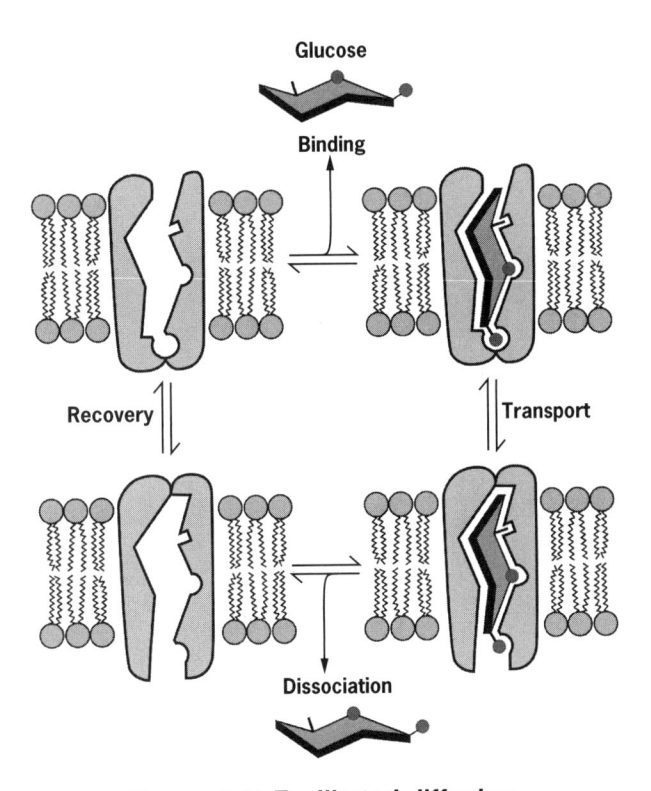

Figure 4.41 Facilitated diffusion.

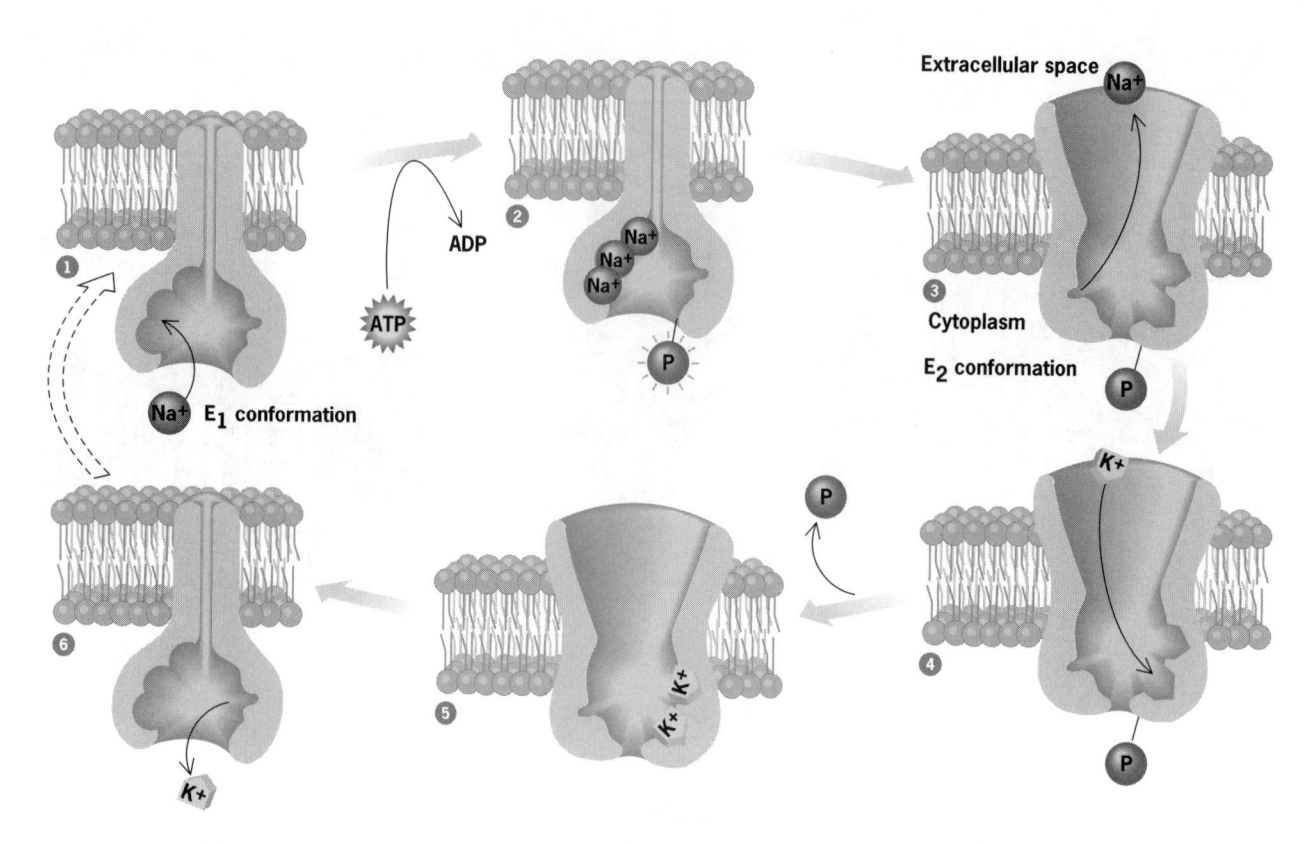

Figure 4.44 Schematic concept of the Na$^+$/K$^+$-ATPase transport cycle.

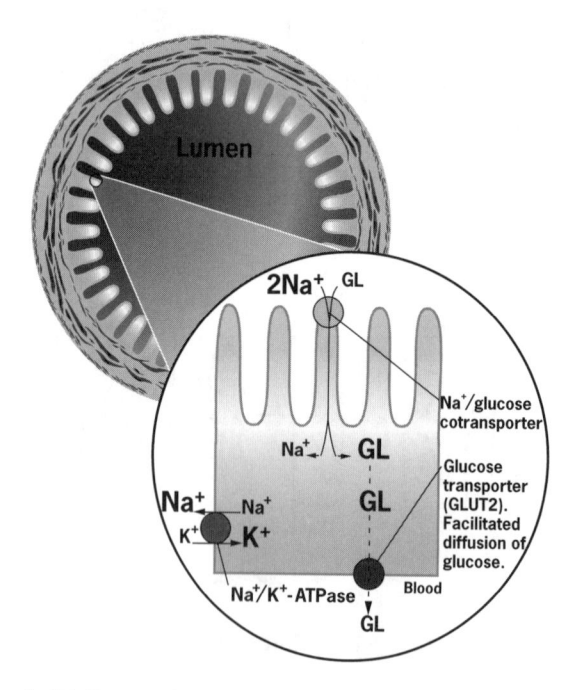

Figure 4.47 Secondary transport: the use of energy stored in an ionic gradient.

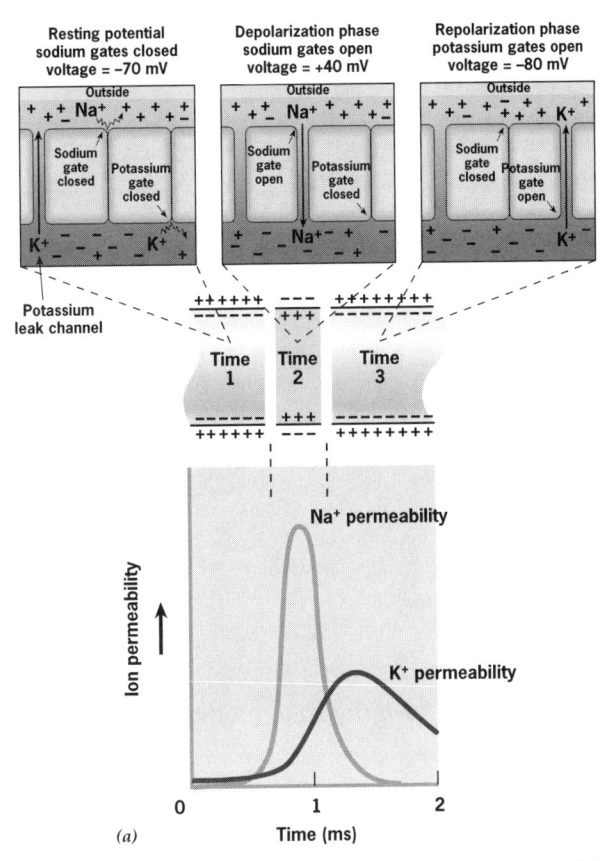

Figure 4.50a Formation of an action potential.

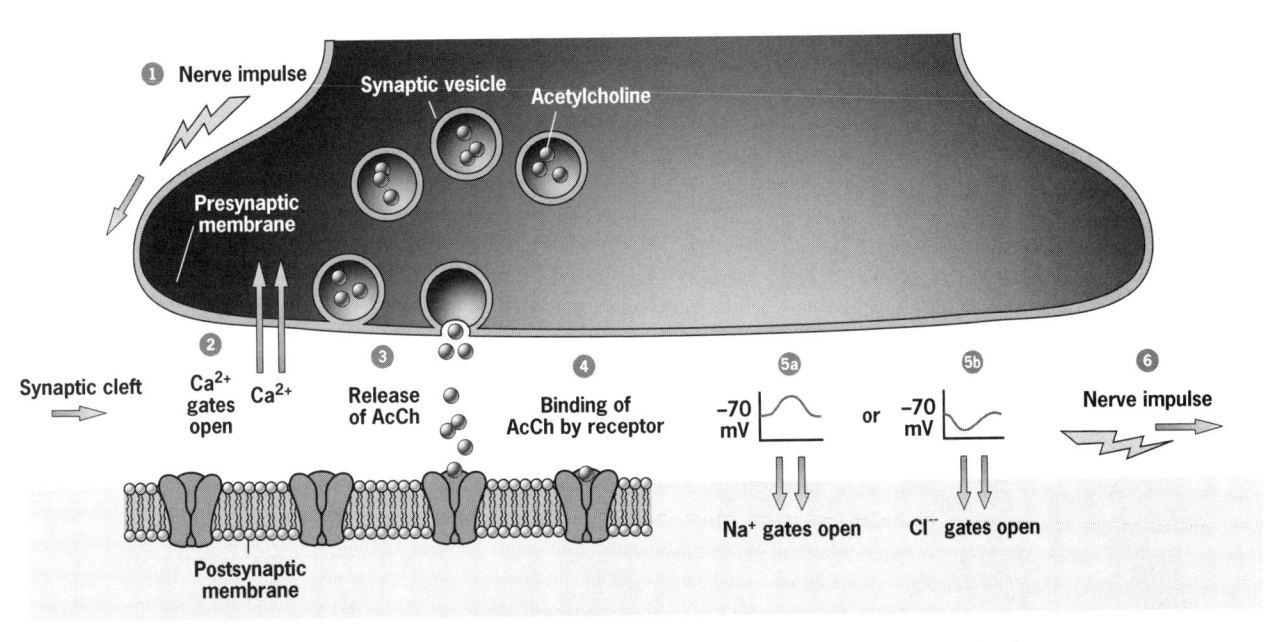

Figure 4.54 The sequence of events during synaptic transmission with acetylcholine as the neurotransmitter.

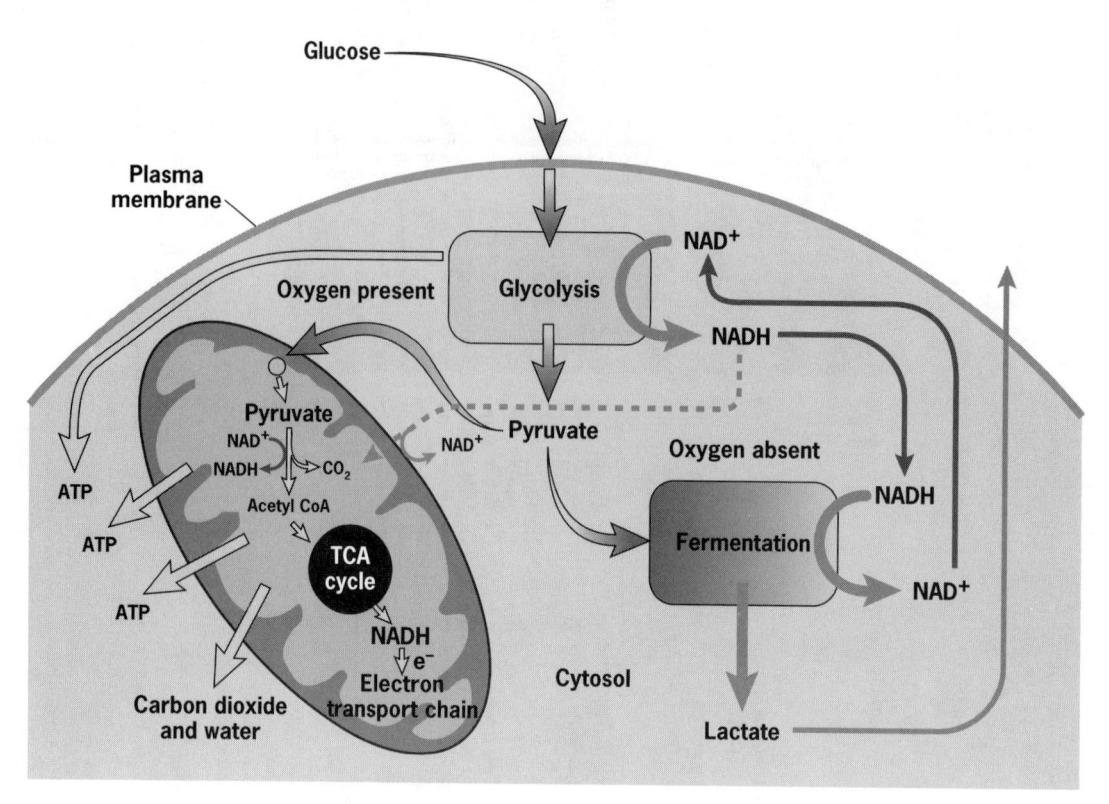

Figure 5.5 An overview of carbohydrate metabolism in eukaryotic cells.

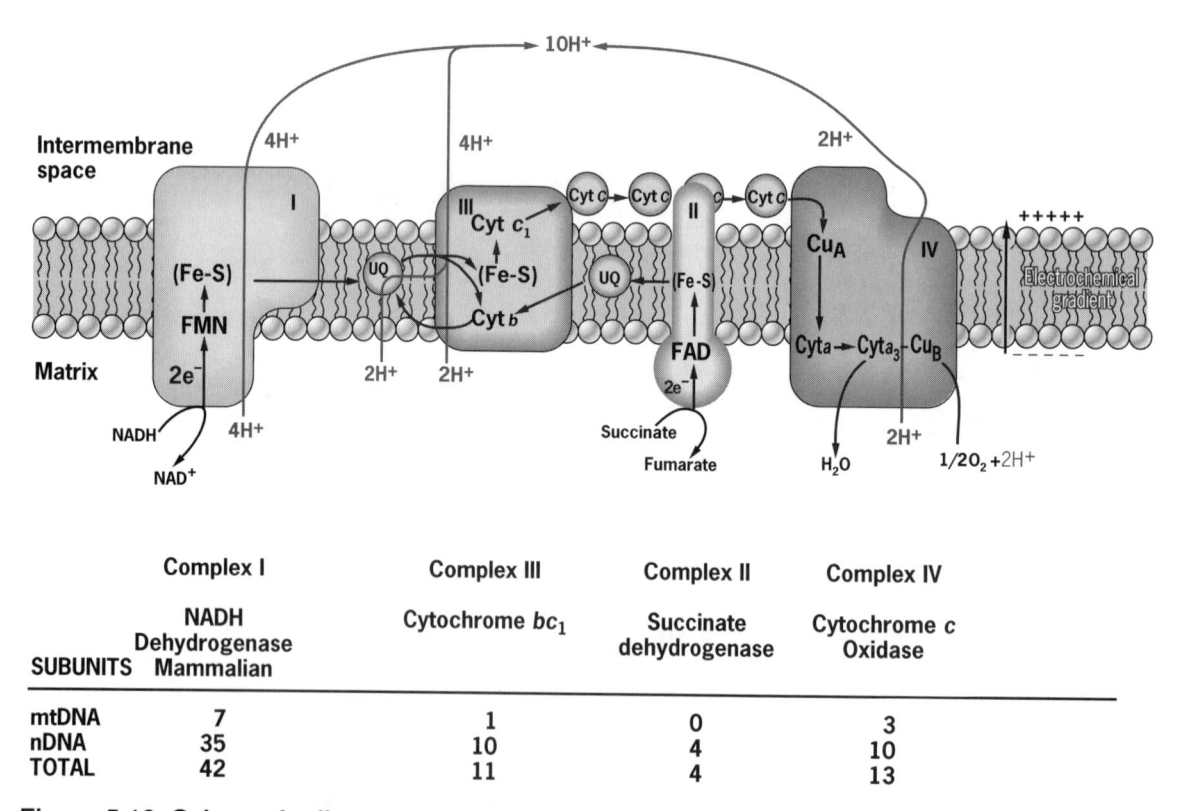

	Complex I	Complex III	Complex II	Complex IV
SUBUNITS	NADH Dehydrogenase Mammalian	Cytochrome bc_1	Succinate dehydrogenase	Cytochrome c Oxidase
mtDNA	7	1	0	3
nDNA	35	10	4	10
TOTAL	42	11	4	13

Figure 5.16 Schematic diagram of the components of the electron-transport chain within the inner mitochondrial membrane.

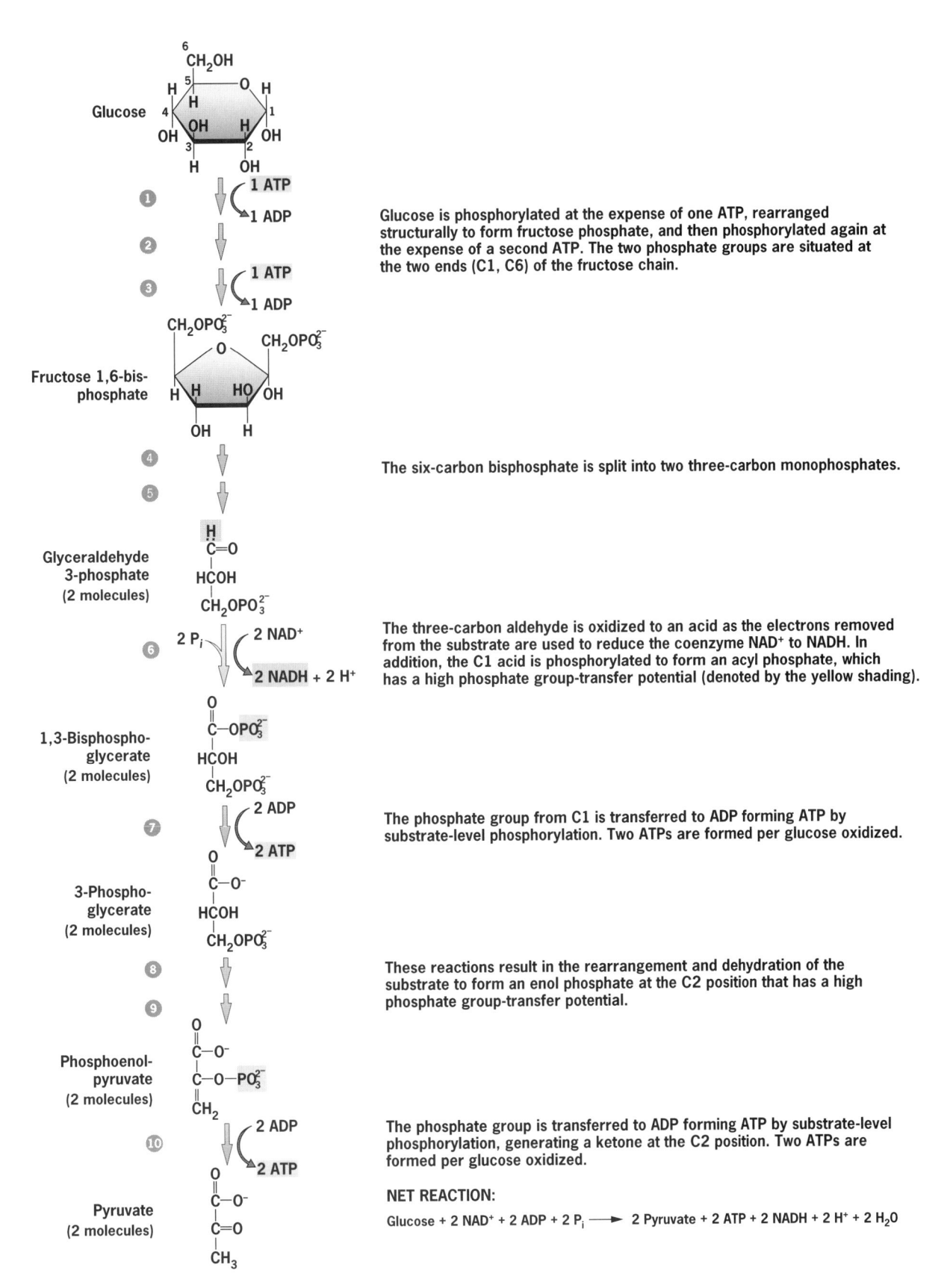

Glucose is phosphorylated at the expense of one ATP, rearranged structurally to form fructose phosphate, and then phosphorylated again at the expense of a second ATP. The two phosphate groups are situated at the two ends (C1, C6) of the fructose chain.

The six-carbon bisphosphate is split into two three-carbon monophosphates.

The three-carbon aldehyde is oxidized to an acid as the electrons removed from the substrate are used to reduce the coenzyme NAD$^+$ to NADH. In addition, the C1 acid is phosphorylated to form an acyl phosphate, which has a high phosphate group-transfer potential (denoted by the yellow shading).

The phosphate group from C1 is transferred to ADP forming ATP by substrate-level phosphorylation. Two ATPs are formed per glucose oxidized.

These reactions result in the rearrangement and dehydration of the substrate to form an enol phosphate at the C2 position that has a high phosphate group-transfer potential.

The phosphate group is transferred to ADP forming ATP by substrate-level phosphorylation, generating a ketone at the C2 position. Two ATPs are formed per glucose oxidized.

NET REACTION:

Glucose + 2 NAD$^+$ + 2 ADP + 2 P$_i$ ⟶ 2 Pyruvate + 2 ATP + 2 NADH + 2 H$^+$ + 2 H$_2$O

Figure 5.6 An overview of glycolysis showing some of the key steps.

Copyright © 2002 John Wiley & Sons, Inc.

Figure 5.6

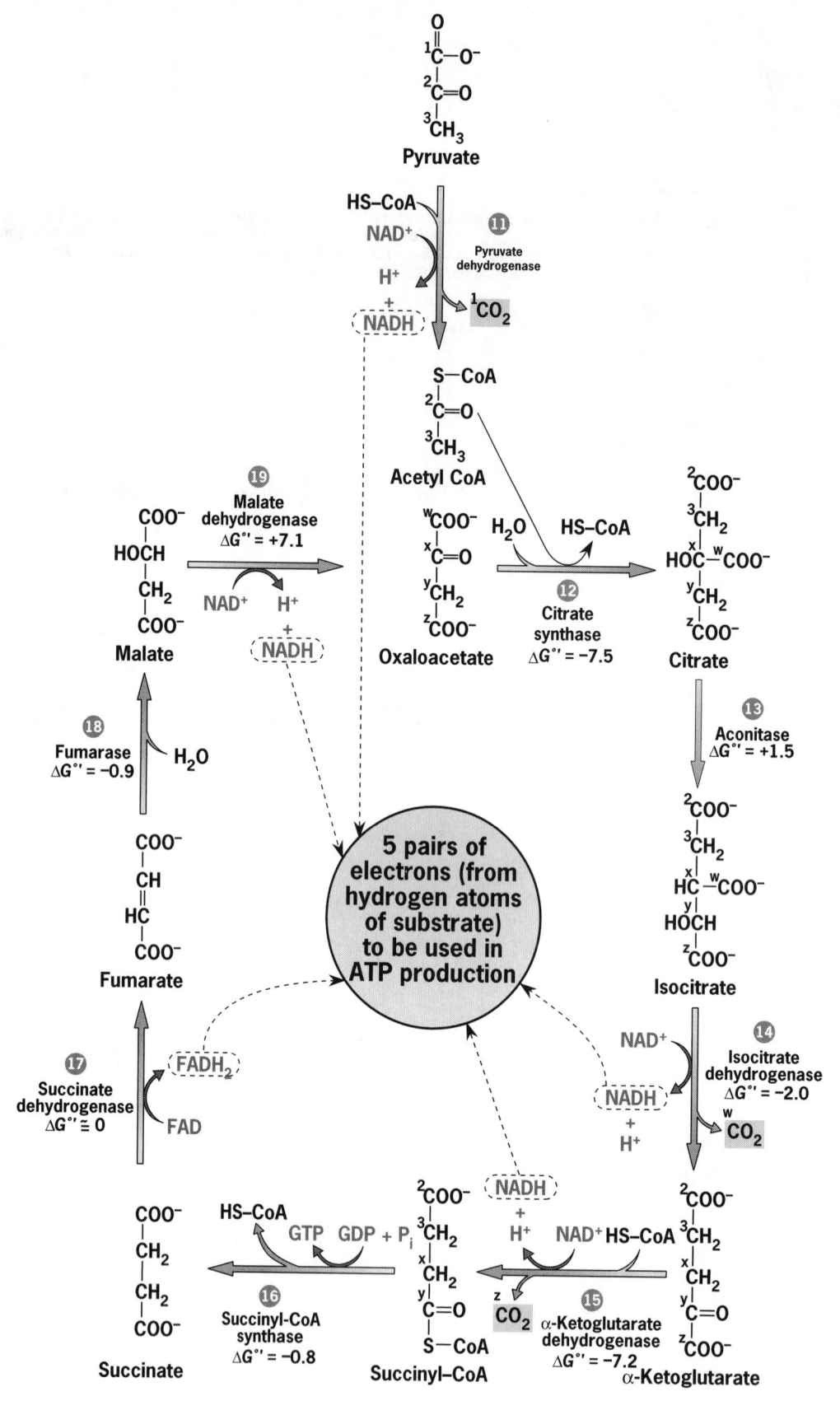

Figure 5.7 The tricarboxylic acid (TCA) cycle.

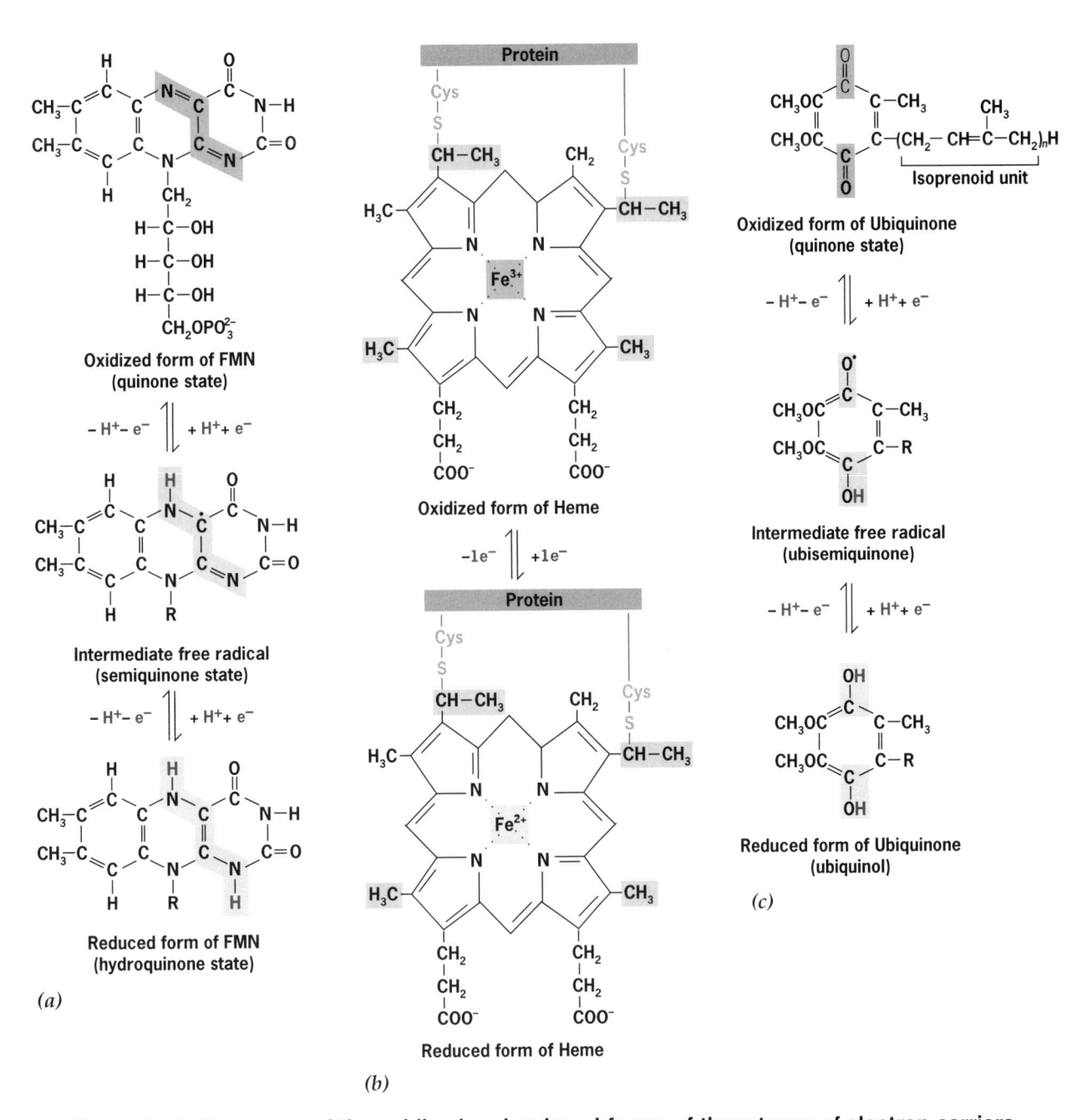

Figure 5.11 Structures of the oxidized and reduced forms of three types of electron carriers.

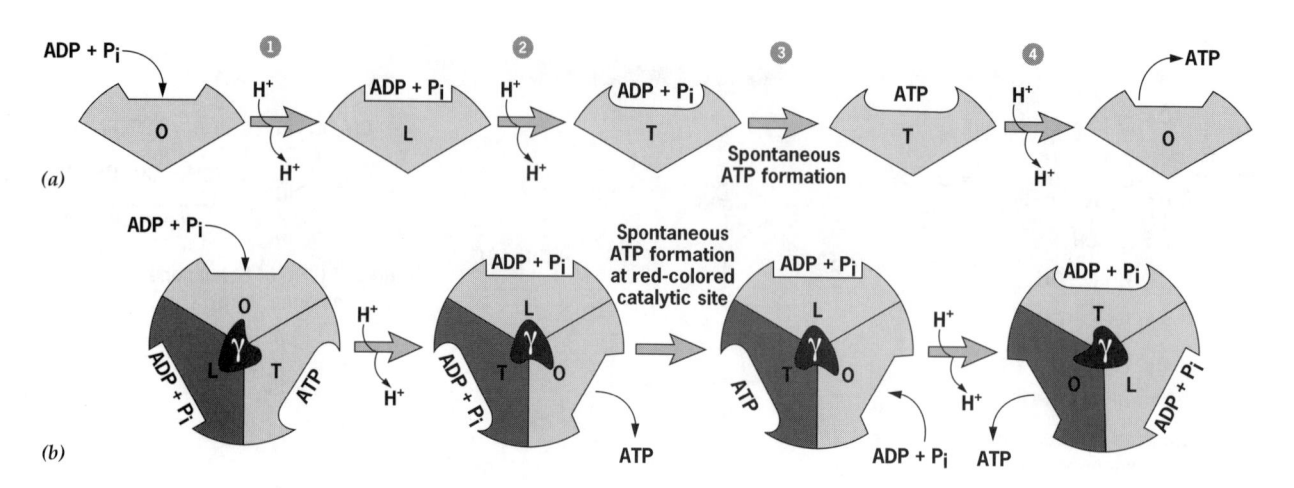

Figure 5.26 The binding change mechanism for ATP synthesis.

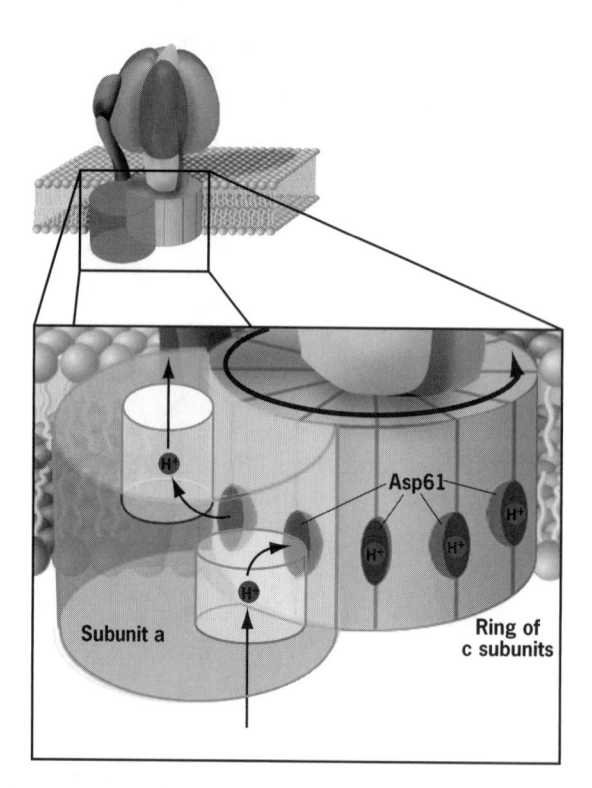

Figure 5.29 A model in which proton diffusion is coupled to the rotation of the c ring of the F_0 complex.

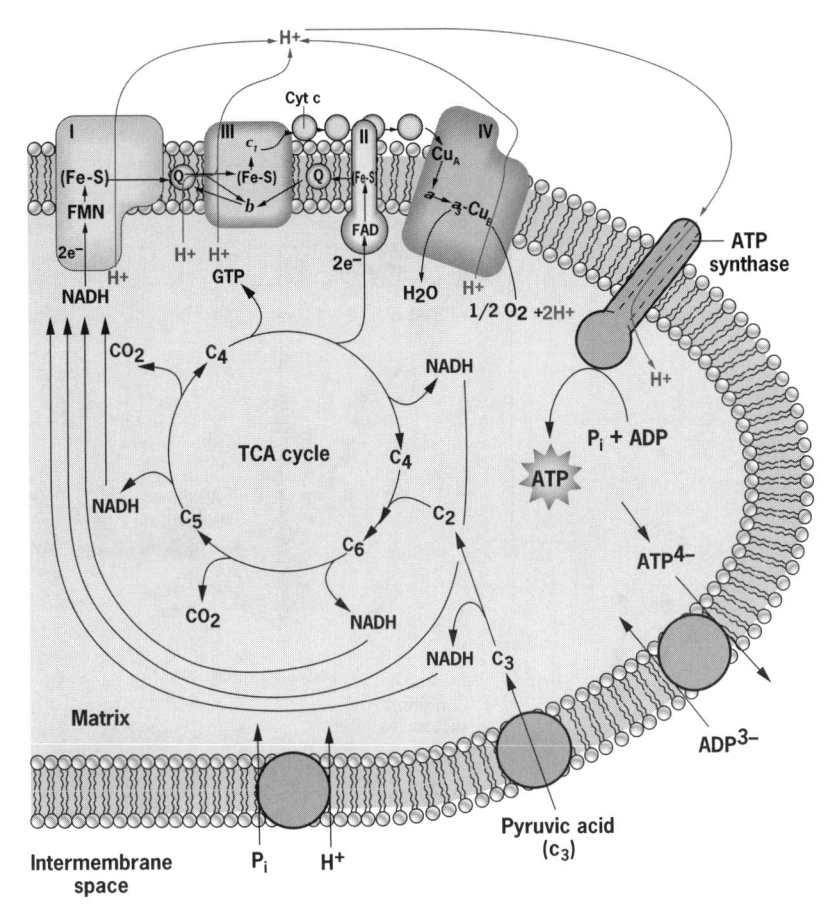

Figure 5.30 Summary of the major activities during aerobic respiration in a mitochondrion.

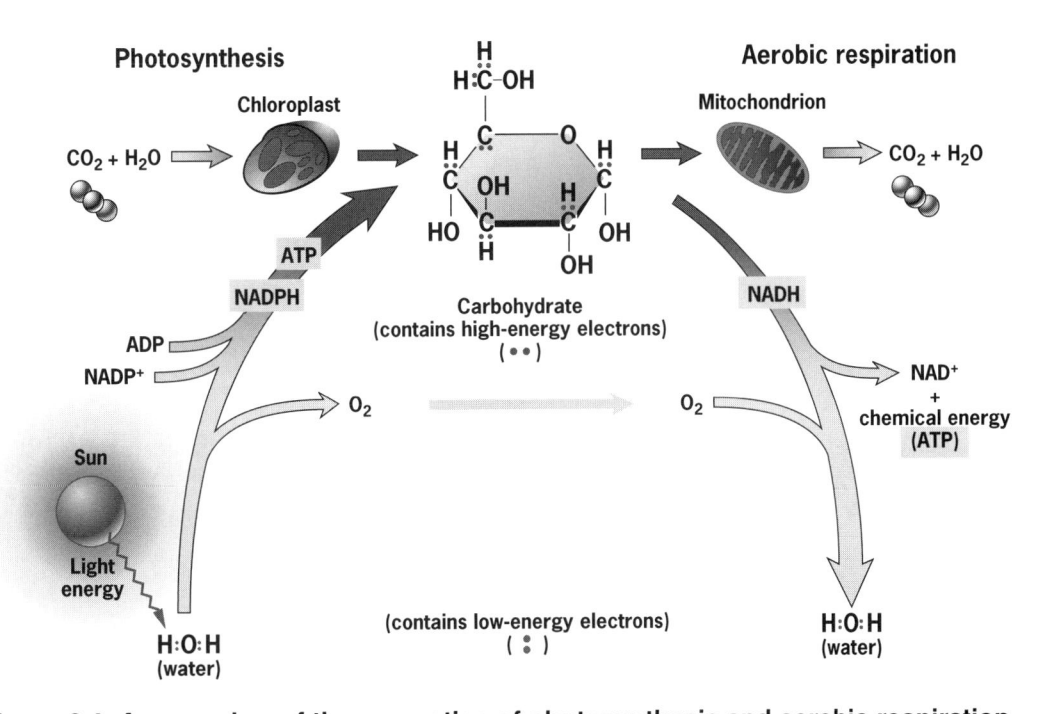

Figure 6.4 An overview of the energetics of photosynthesis and aerobic respiration.

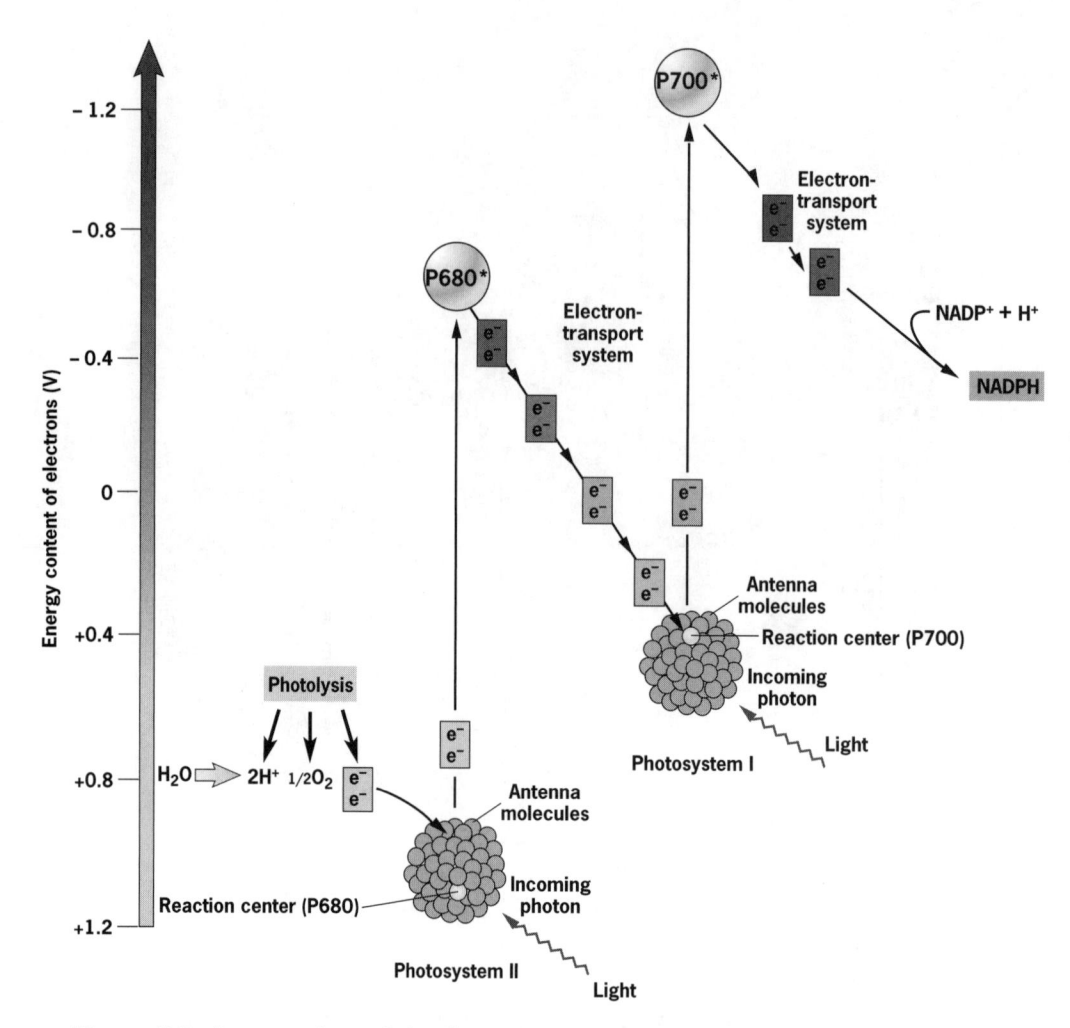

Figure 6.9 An overview of the flow of electrons during the light-dependent reactions of photosynthesis.

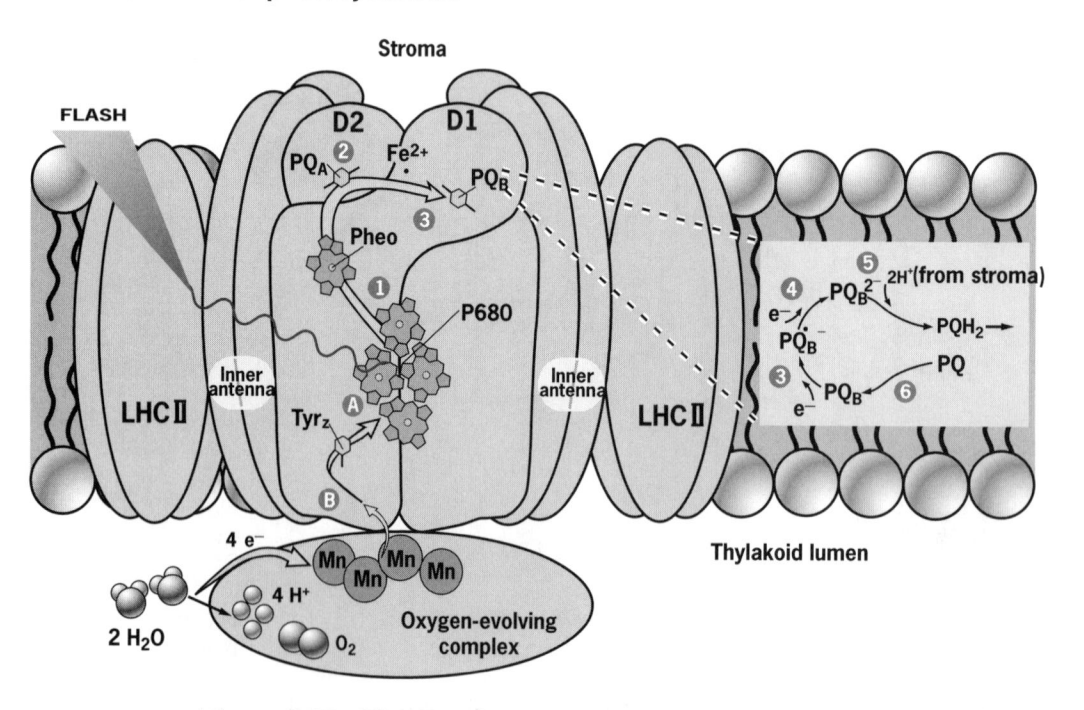

Figure 6.10a The functional organization of photosystem II.

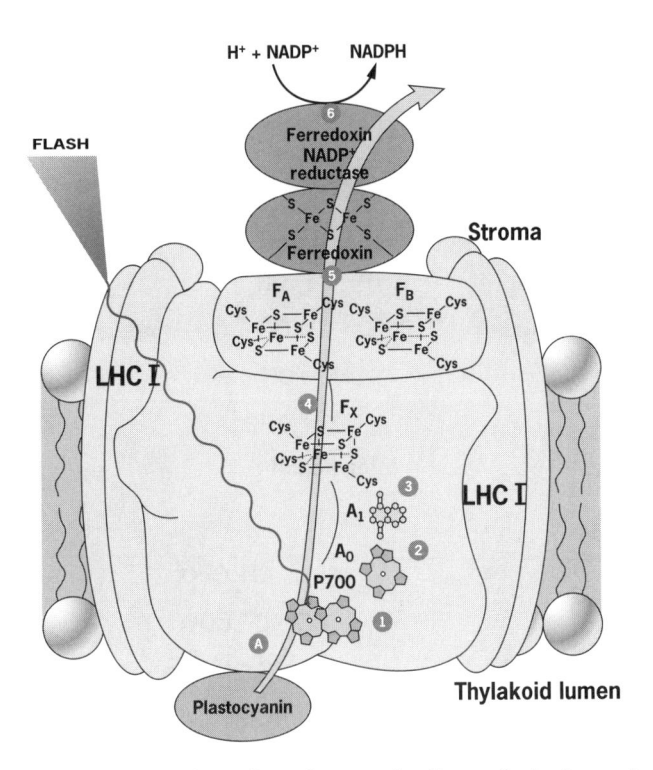

Figure 6.14 The functional organization of photosystem I.

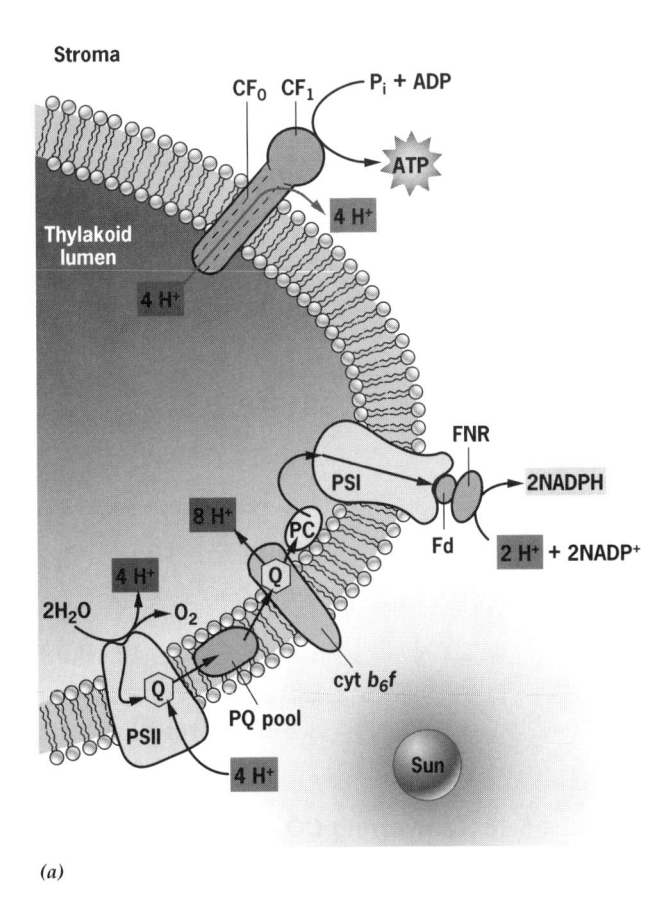

Figure 6.15a Summary of the light-dependent reactions.

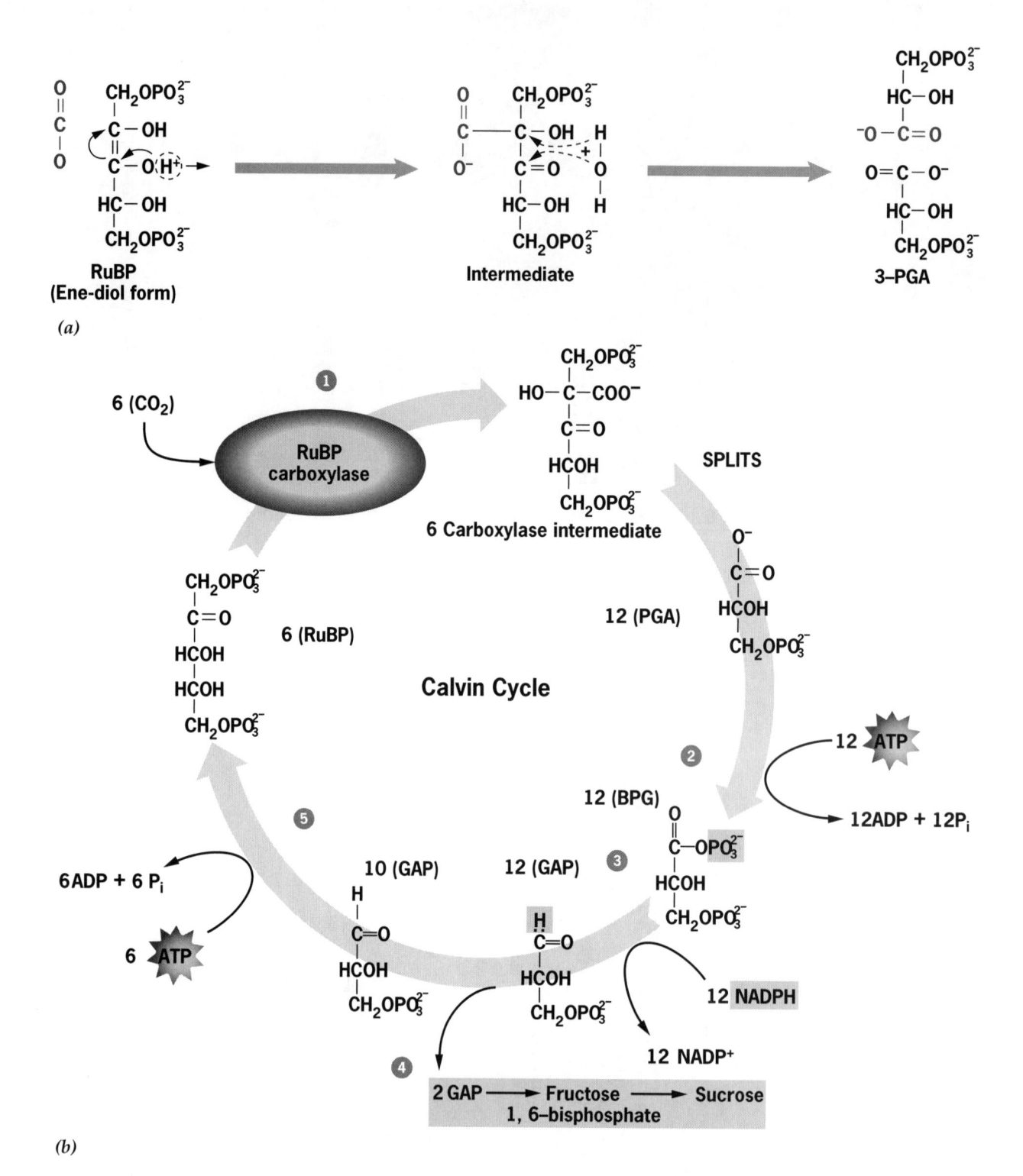

Figure 6.18 Converting CO_2 into carbohydrate.

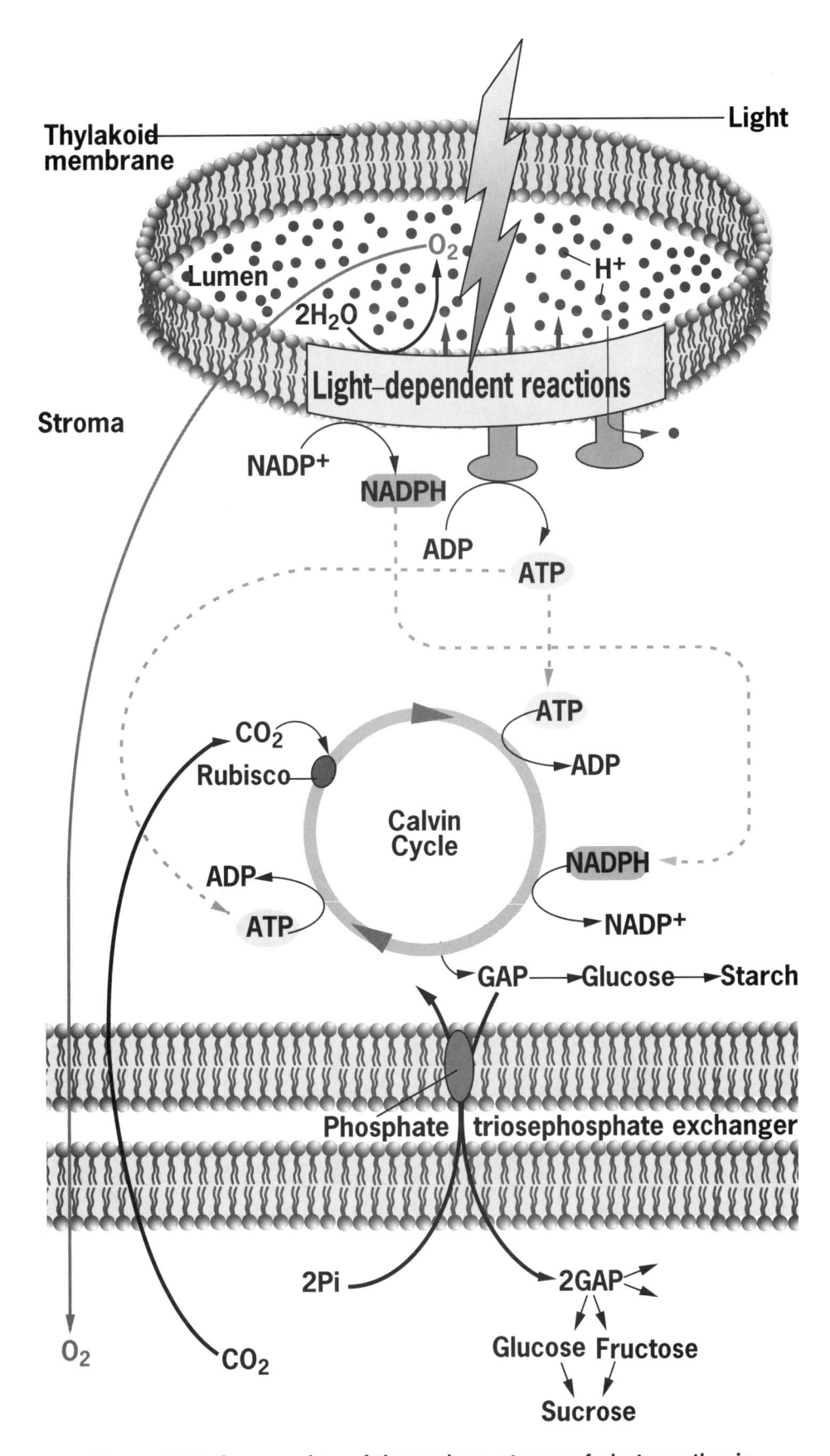

Figure 6.19 An overview of the various stages of photosynthesis.

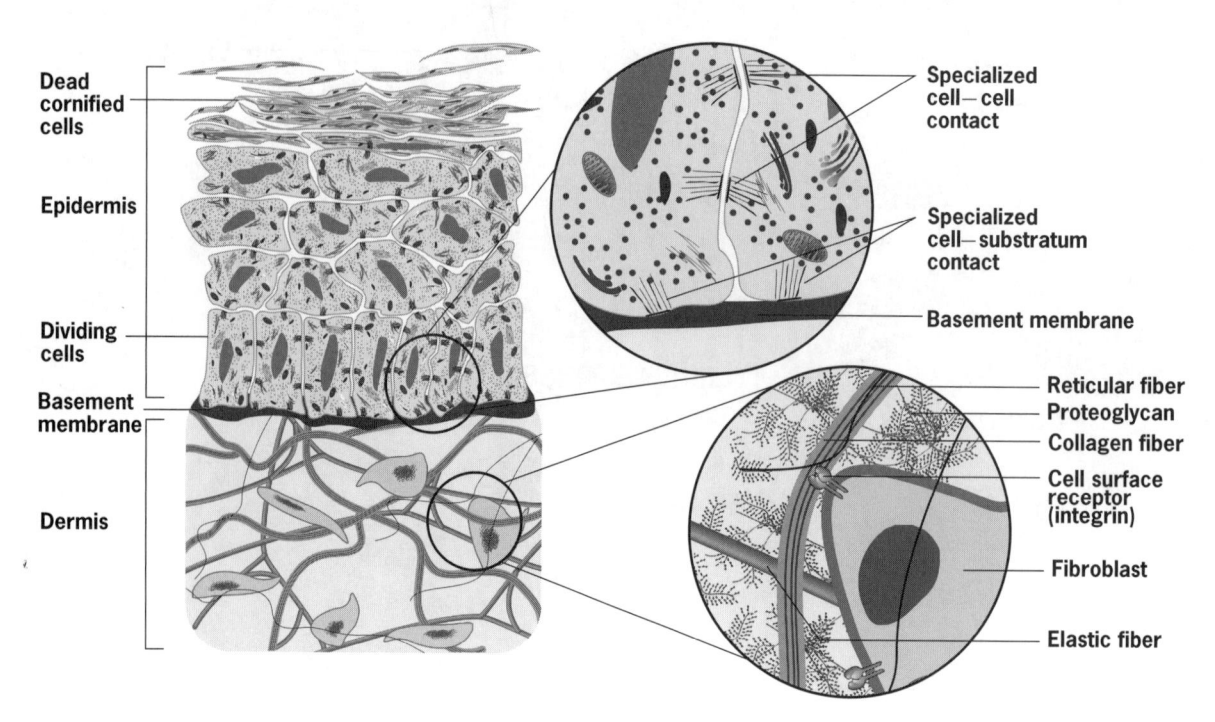

Figure 7.1 An overview of how cells are organized into tissues and how they interact with one another and with their extracellular environment.

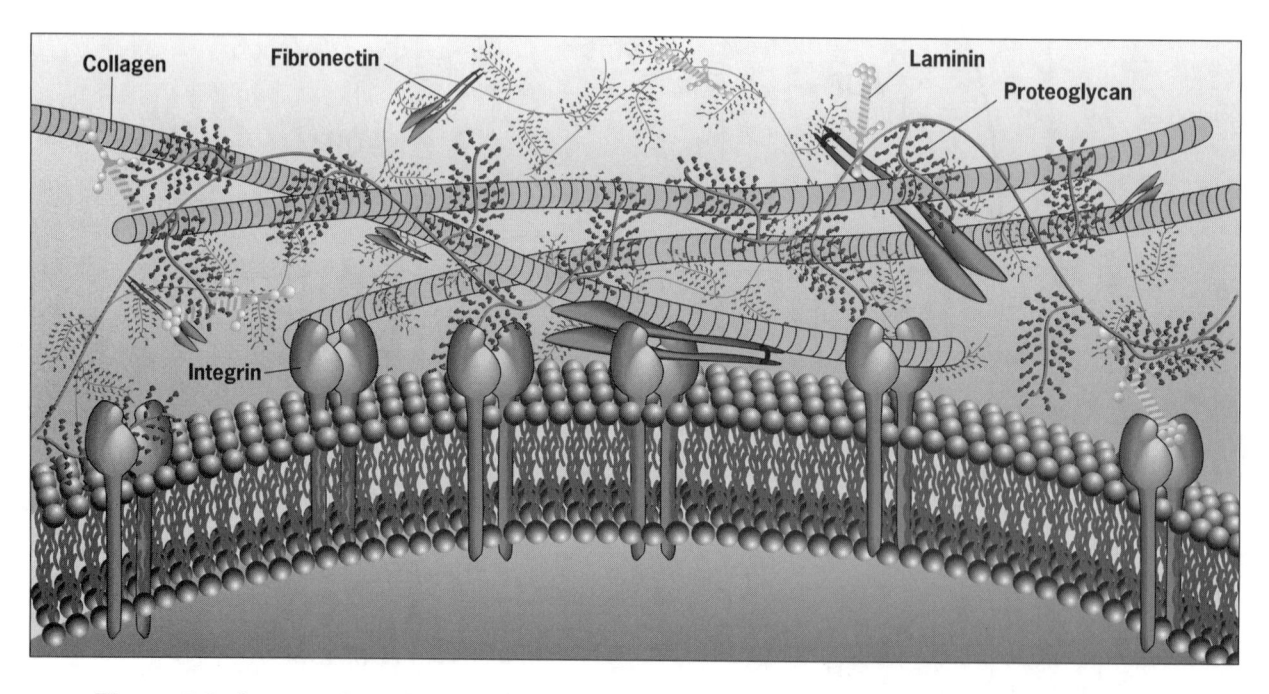

Figure 7.5 An overview of the macromolecular organization of the extracellular matrix.

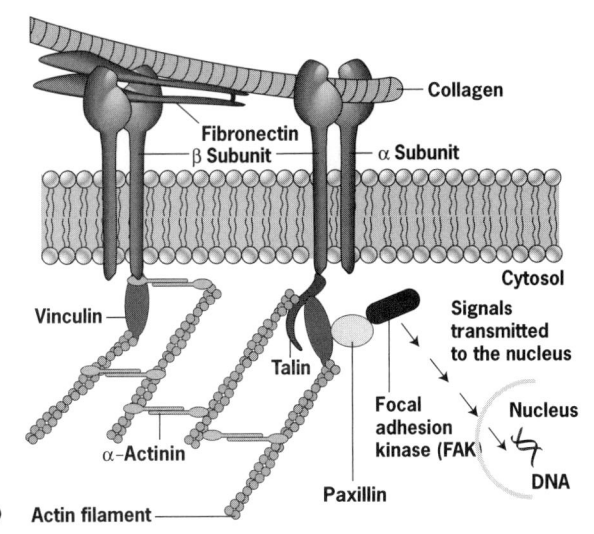

Figure 7.17c Focal adhesions are sites where cells adhere to their substratum.

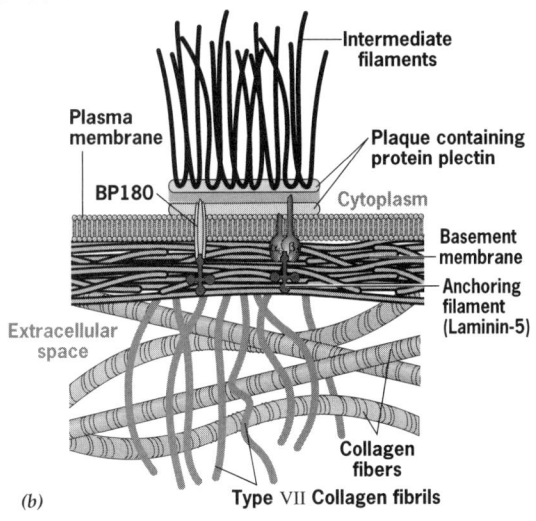

Figure 7.18b Hemidesmosomes.

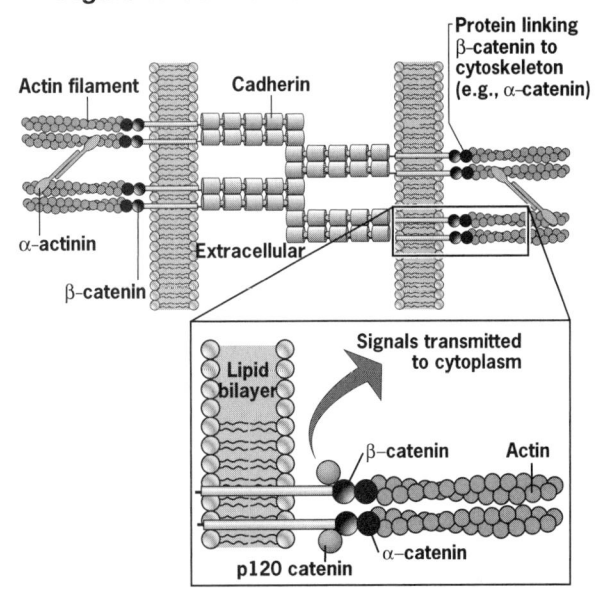

Figure 7.26 The structure of an adherens junction.

Figures 7.17c & 7.18b & 7.26

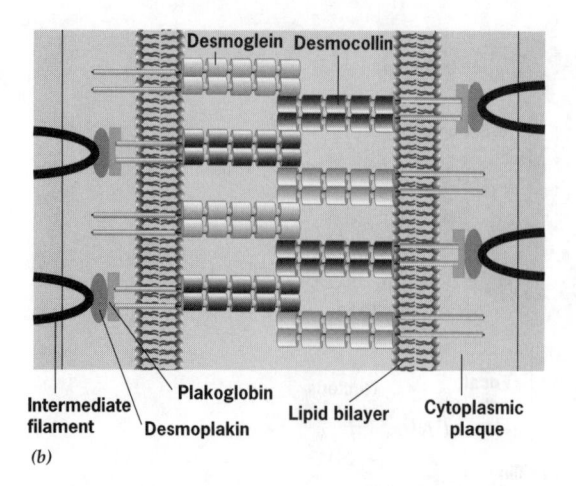

Figure 7.27b The structure of a desmosome.

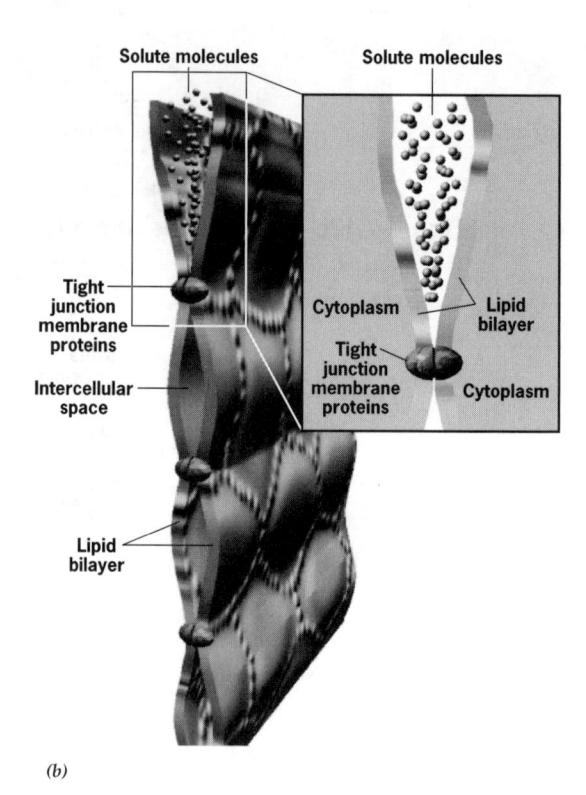

Figure 7.30b Tight junctions.

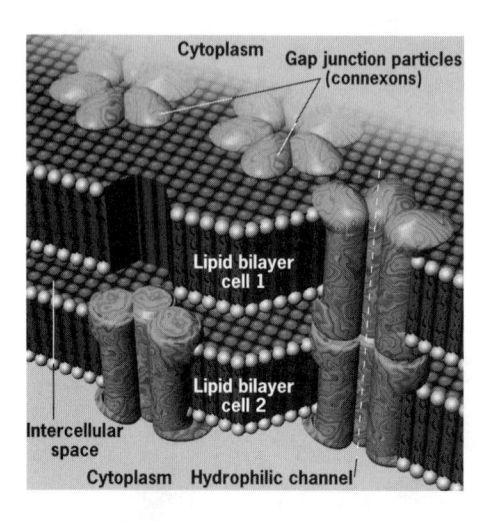

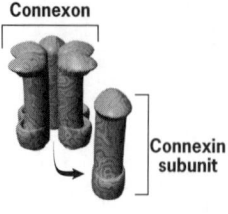

Figure 7.32b The structure of a gap junction.

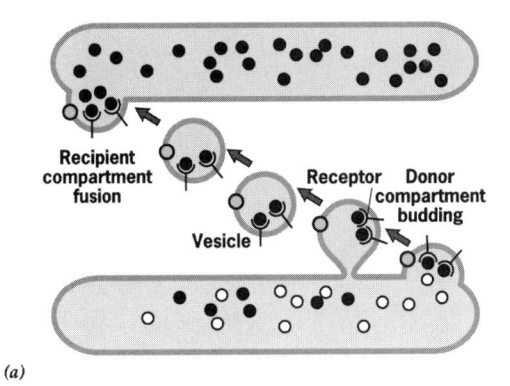

Figure 8.2 An overview of the biosynthetic/secretory and endocytic pathways that unite endomembranes into a dynamic, interconnected network.

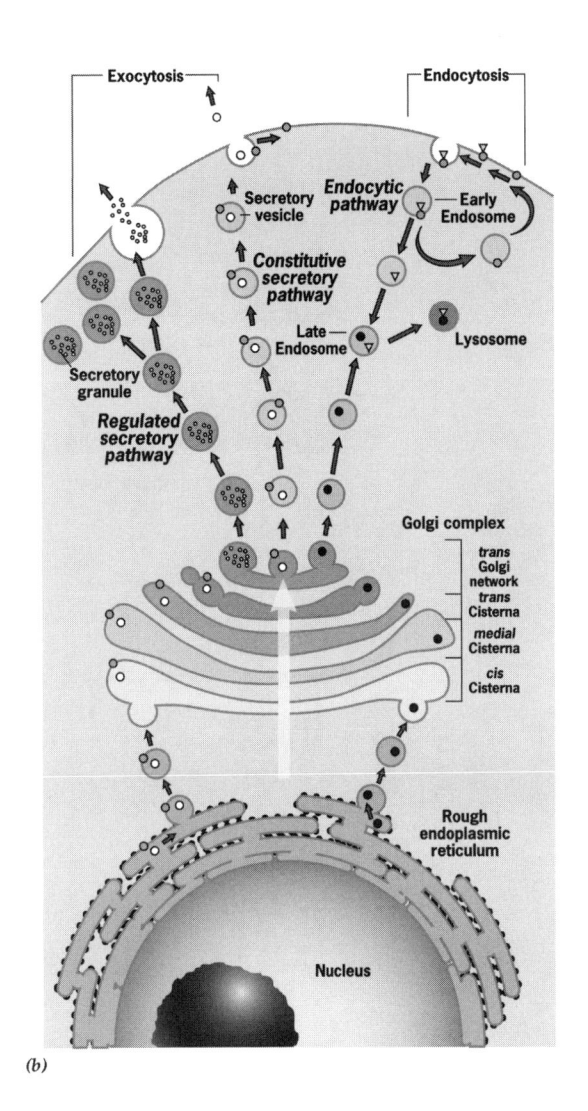

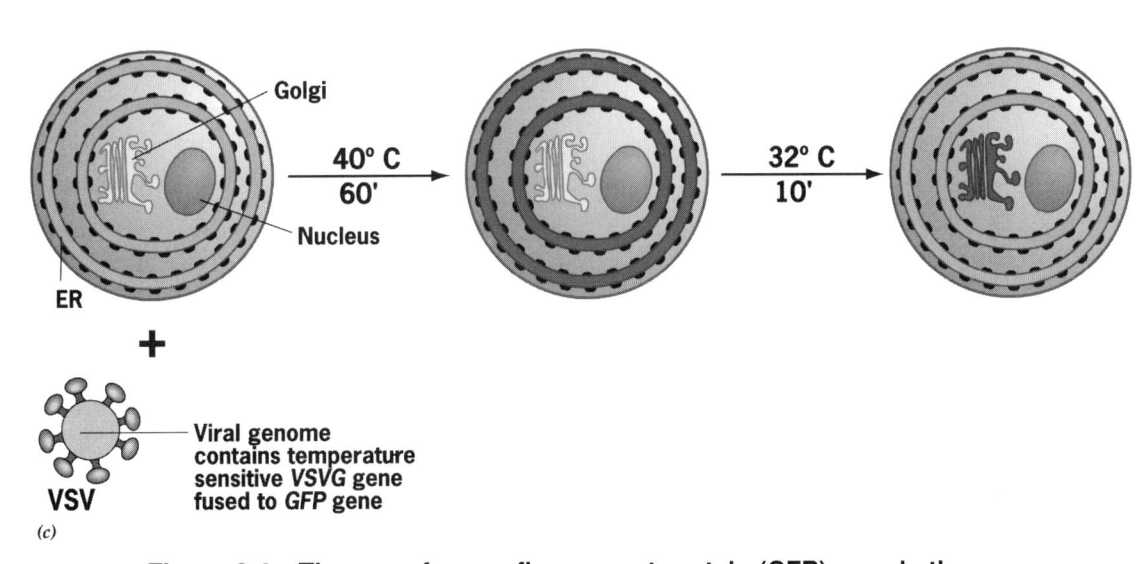

Figure 8.4c The use of green fluorescent protein (GFP) reveals the movement of proteins within a living cell.

Figures 8.2 & 8.4c

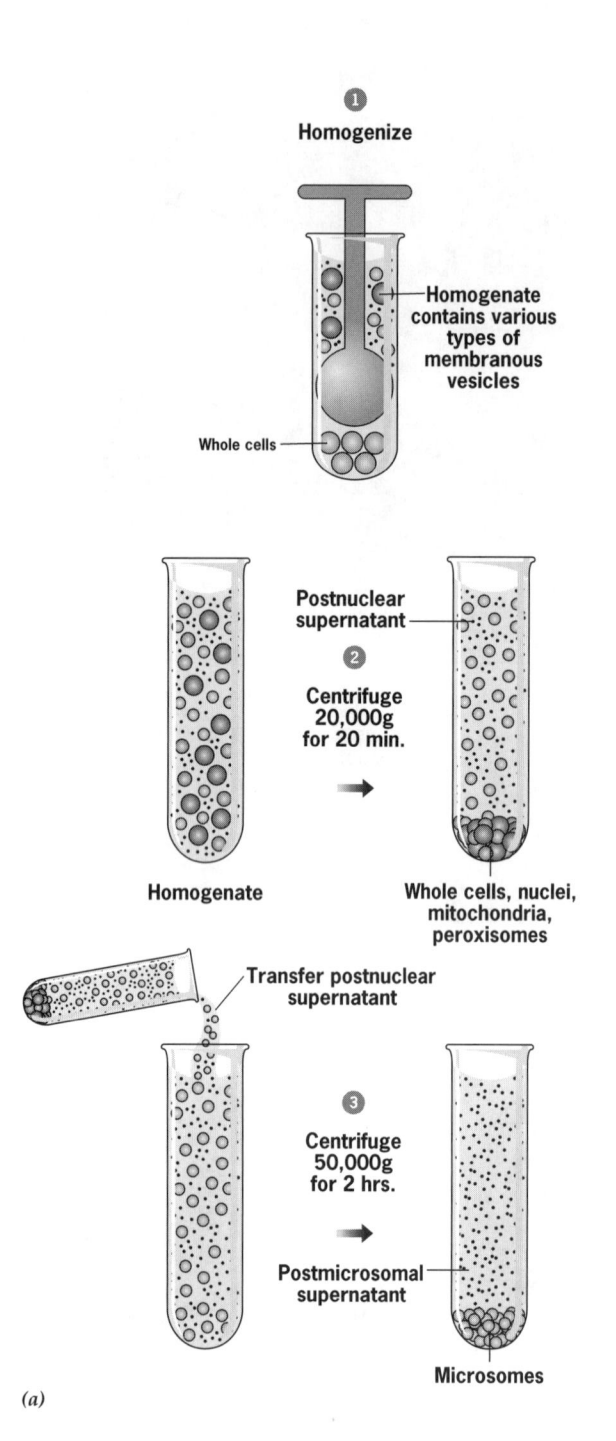

Figure 8.5a Isolation of a microsomal fraction by differential centrifugation.

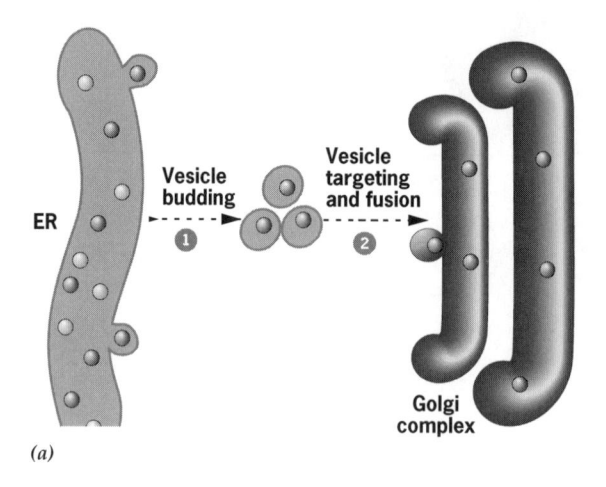

Figure 8.7a The use of genetic mutants in the study of secretion.

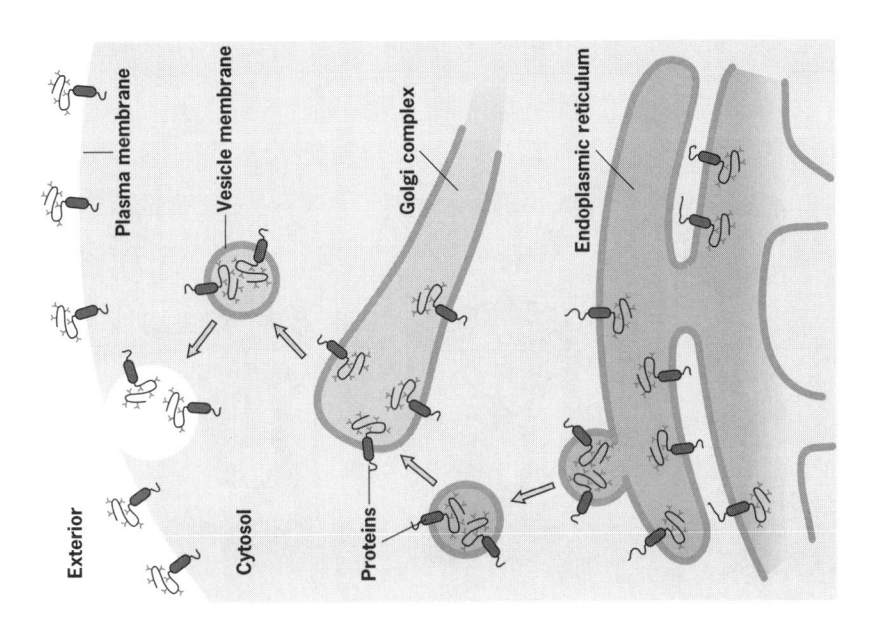

Figure 8.14 Maintenance of membrane asymmetry.

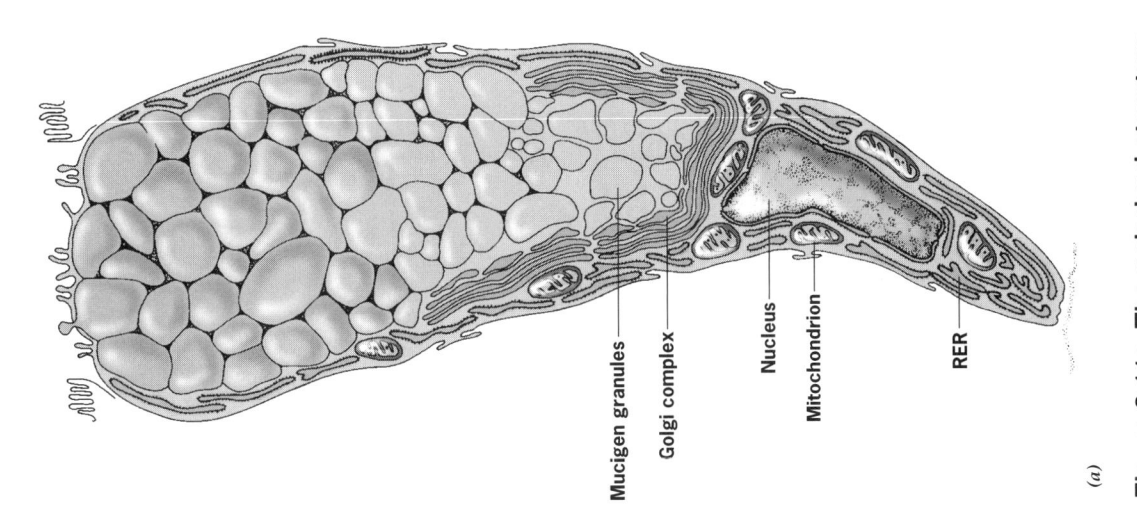

Figure 8.11a The polarized structure of a secretory cell.

Copyright © 2002 John Wiley & Sons, Inc.

Figures 8.11a & 8.14

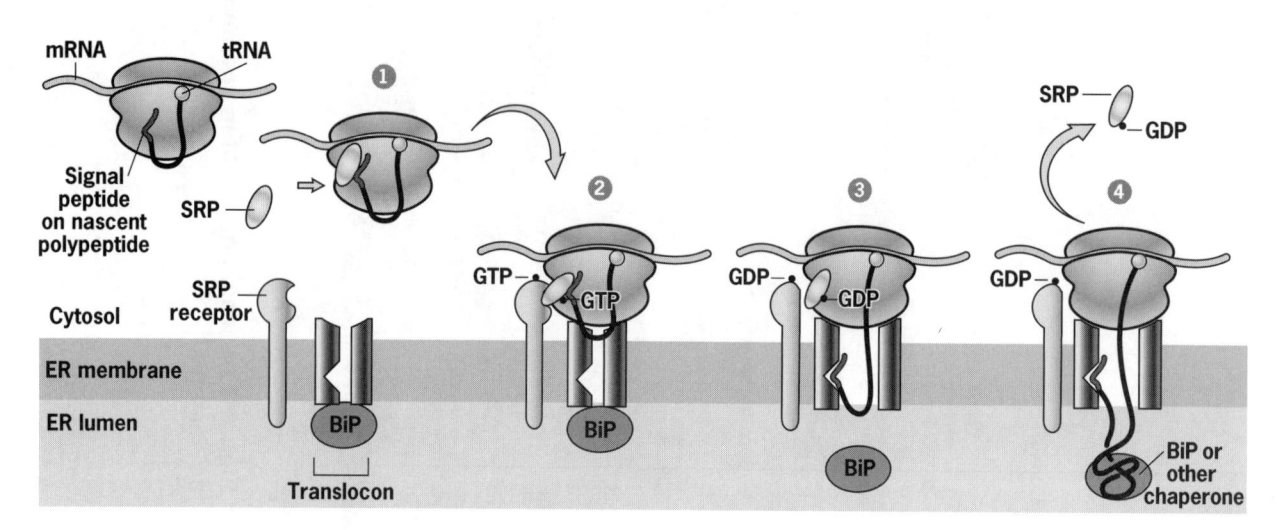

Figure 8.12 A schematic model for the synthesis of a secretory protein (or a lysosomal enzyme) on a membrane-bound ribosome of the rough ER.

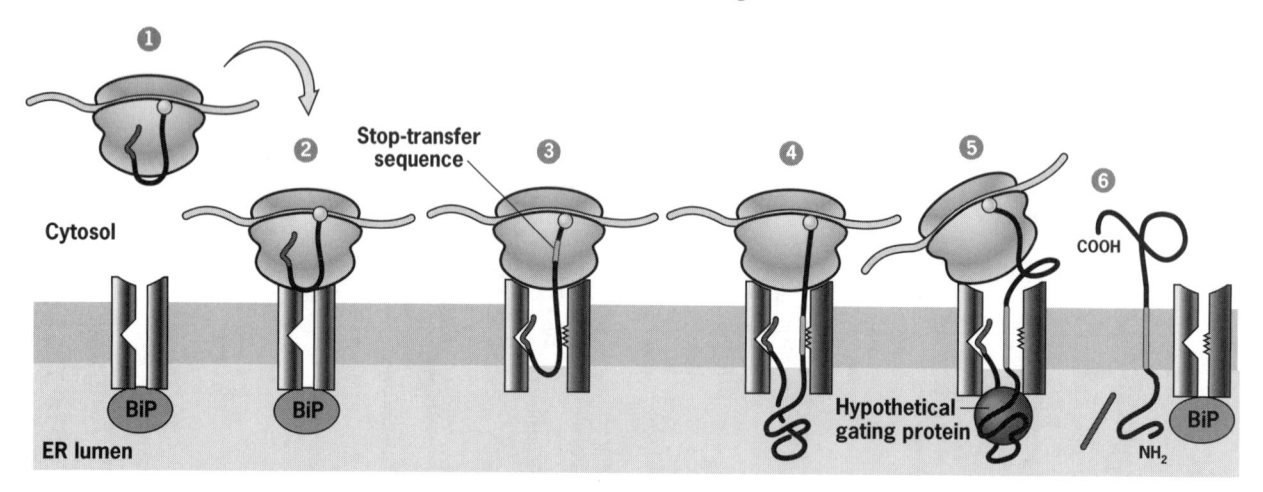

Figure 8.13 A model for the synthesis of an integral membrane protein.

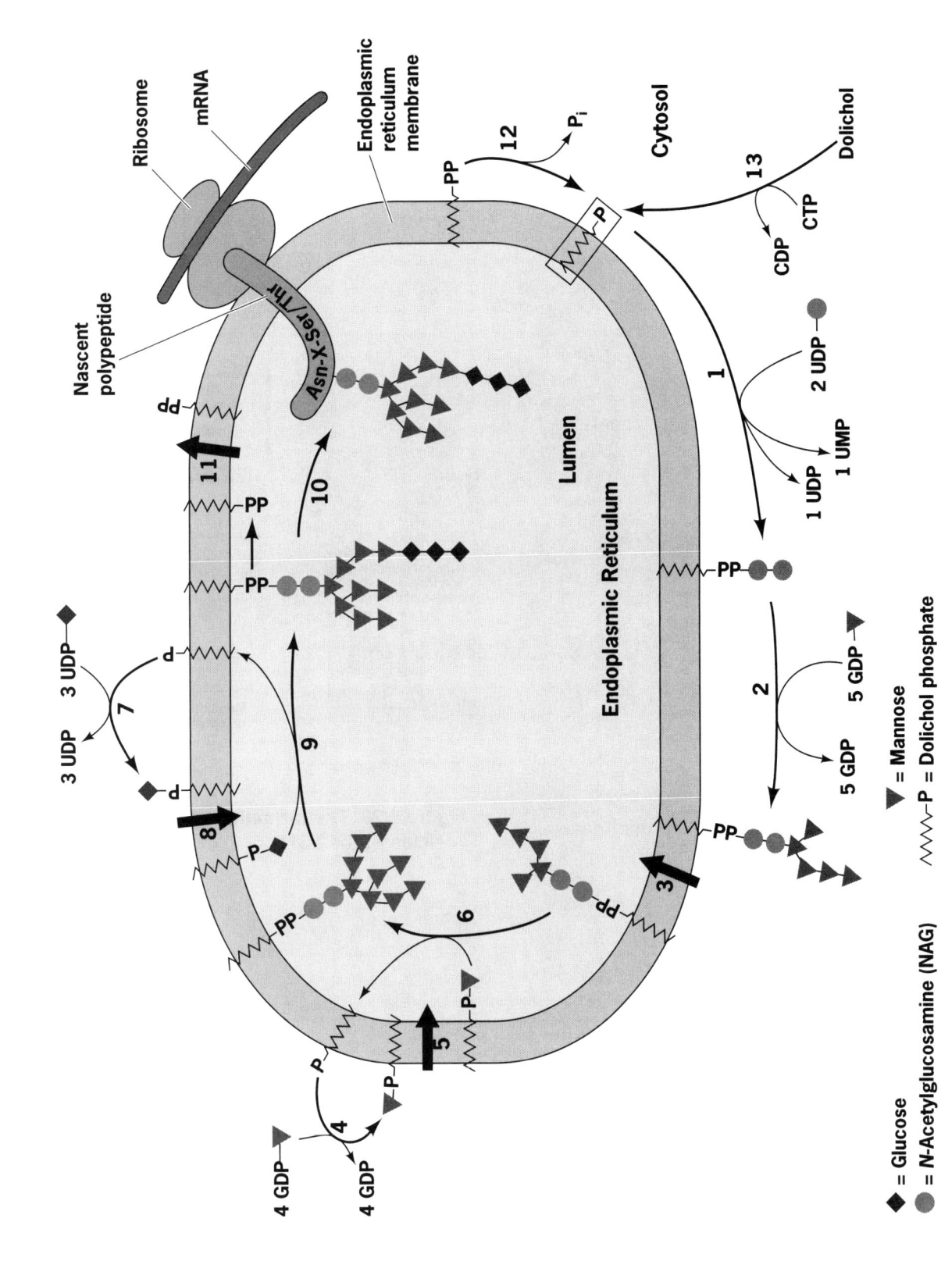

Figure 8.17 The steps in the synthesis of the core portion of an *N*-linked oligosaccharide in the rough ER.

Figure 8.17

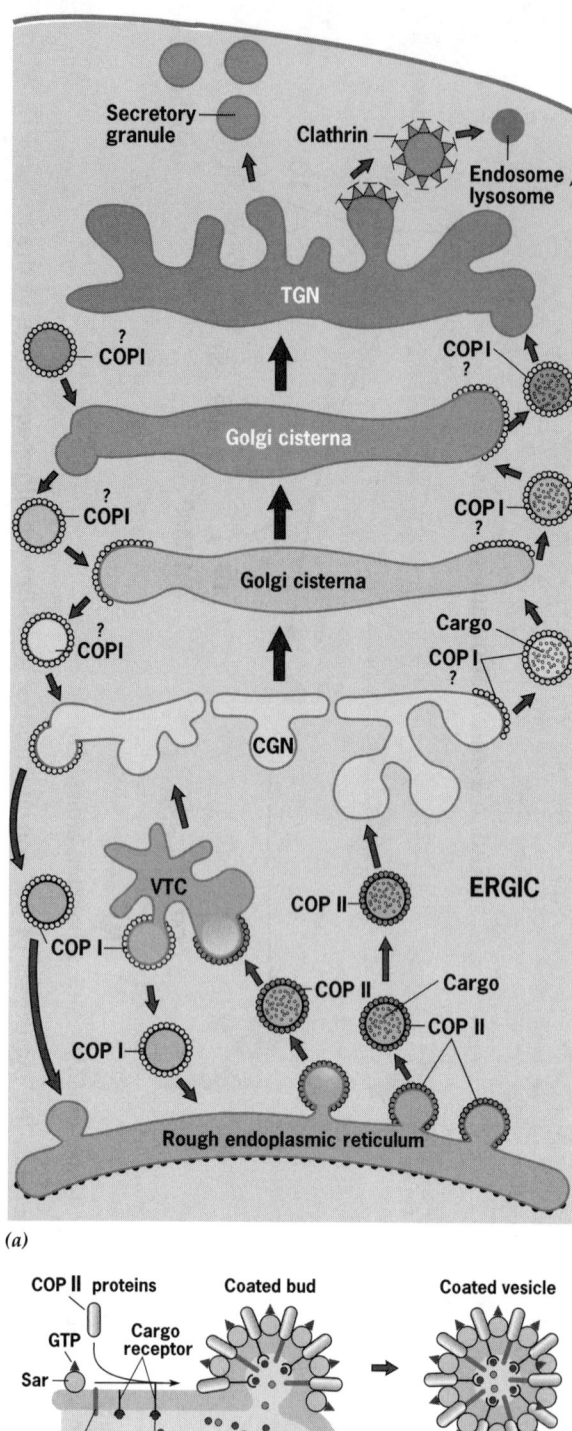

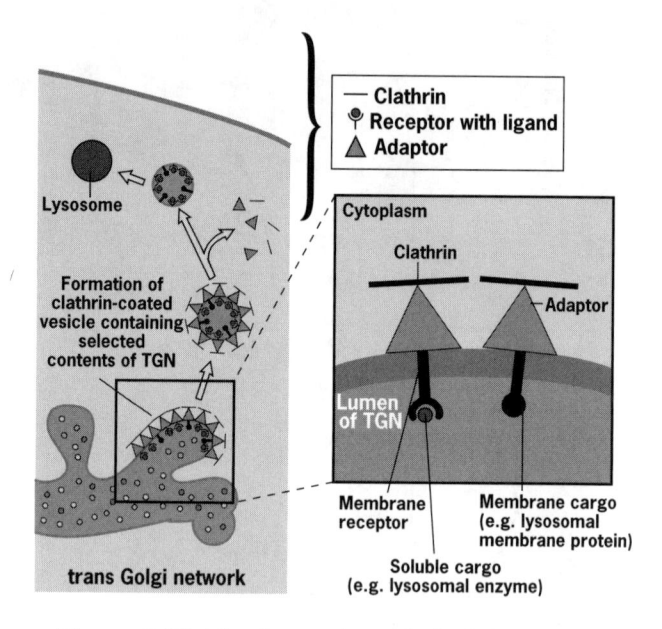

Figure 8.28 The formation of clathrin-coated vesicles at the TGN.

Figure 8.25 Movement of materials by vesicular transport between membranous compartments of the biosynthetic/secretory pathway.

Copyright © 2002 John Wiley & Sons, Inc.

Figures 8.25 & 8.28

Figure 8.29b The mechanism by which lysosomal enzymes are targeted to lysosomes.

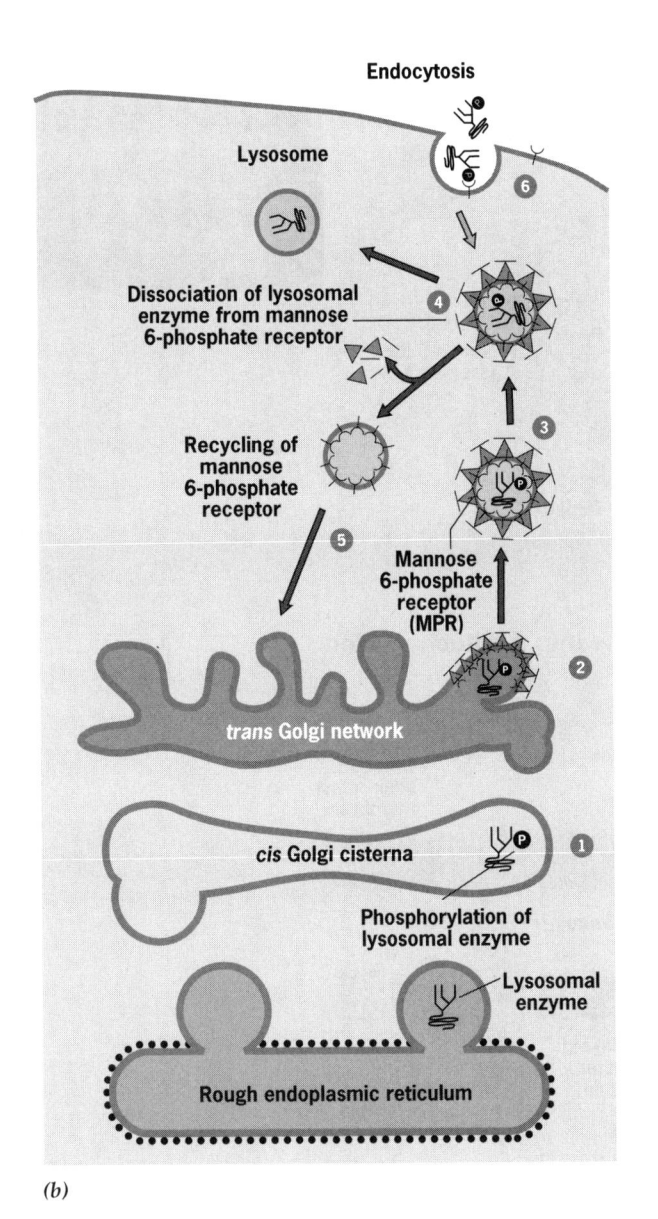

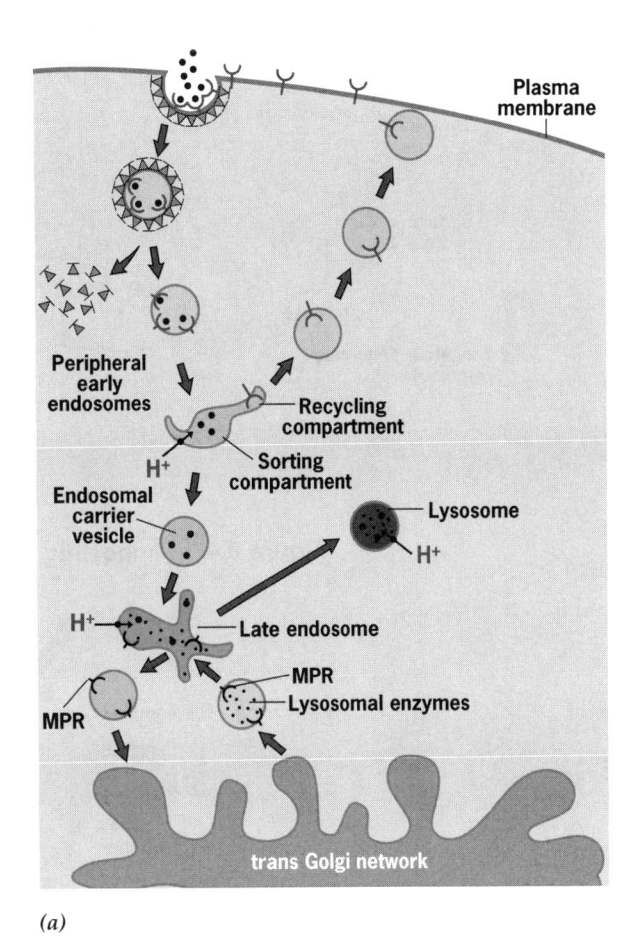

Figure 8.39a The endocytic pathway.

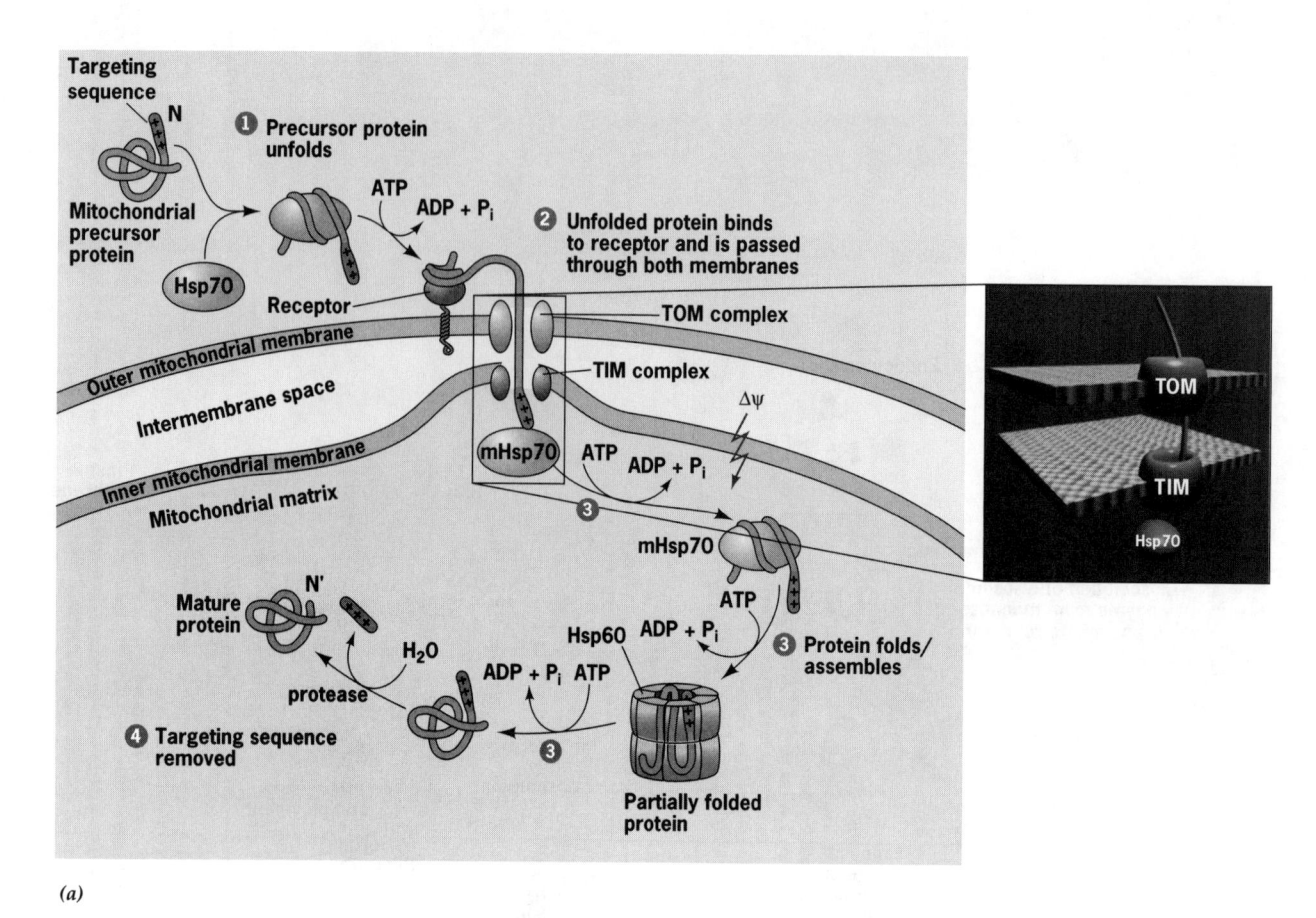

Figure 8.47a Importing proteins into a mitochondrion.

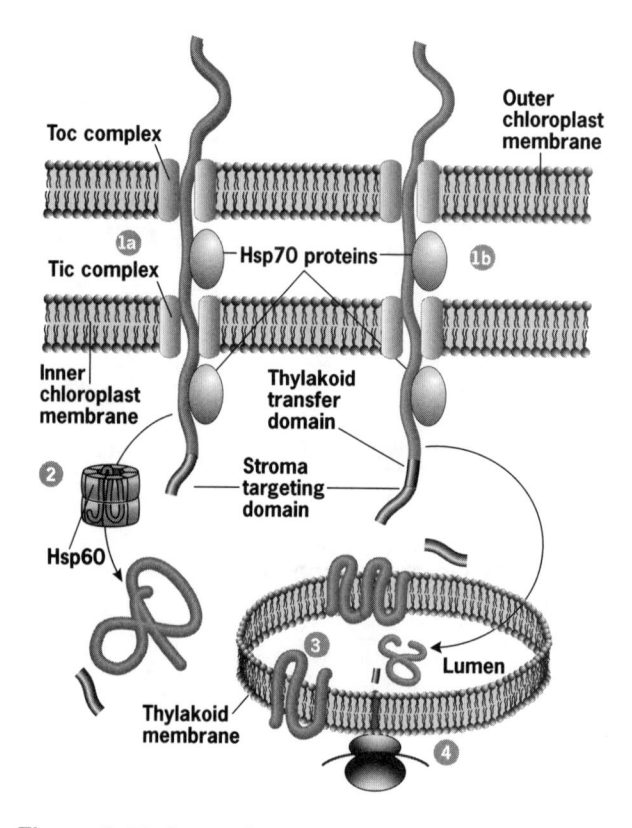

Figure 8.48 Importing proteins into a chloroplast.

Key to Cytoskeletal Functions

(1) Structure and Support (2) Intracellular Transport (3) Contractility and Motility (4) Spatial Organization

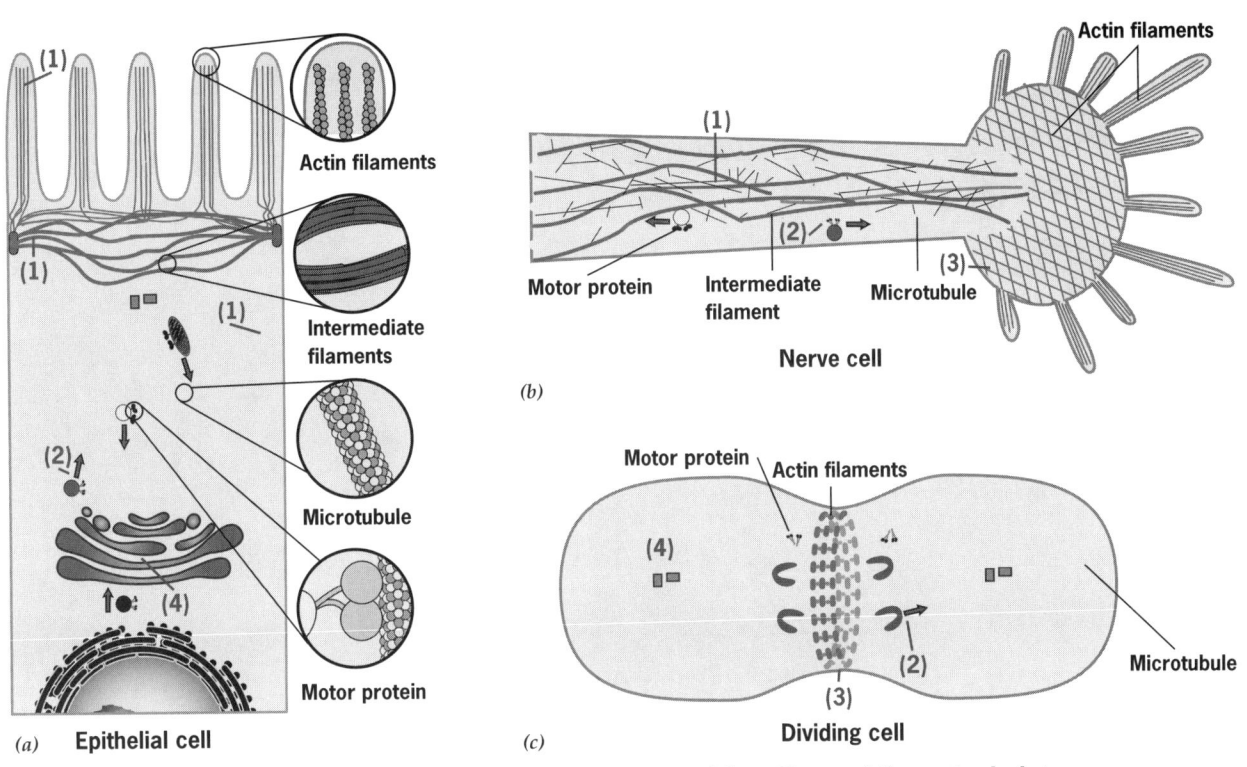

Figure 9.1 An overview of the structure and functions of the cytoskeleton.

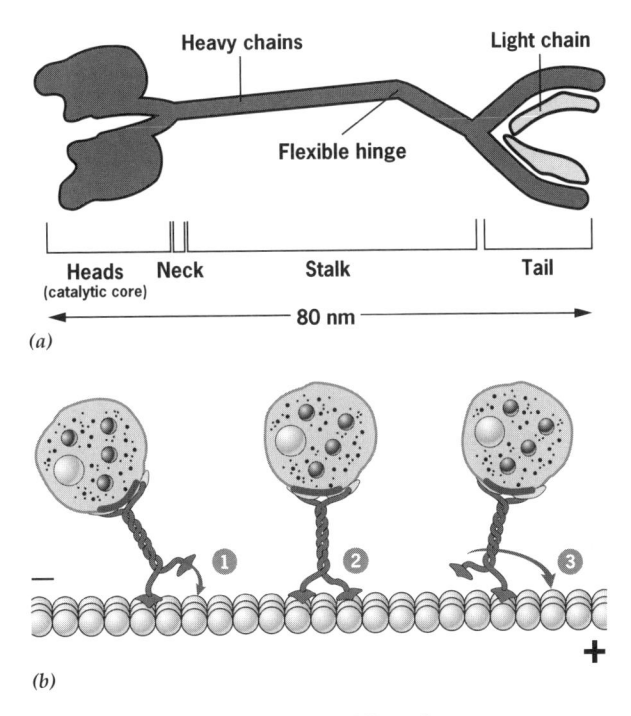

Figure 9.16a,b Kinesin.

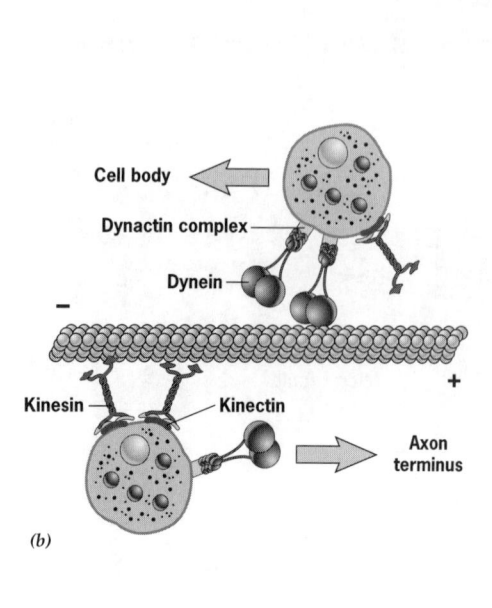

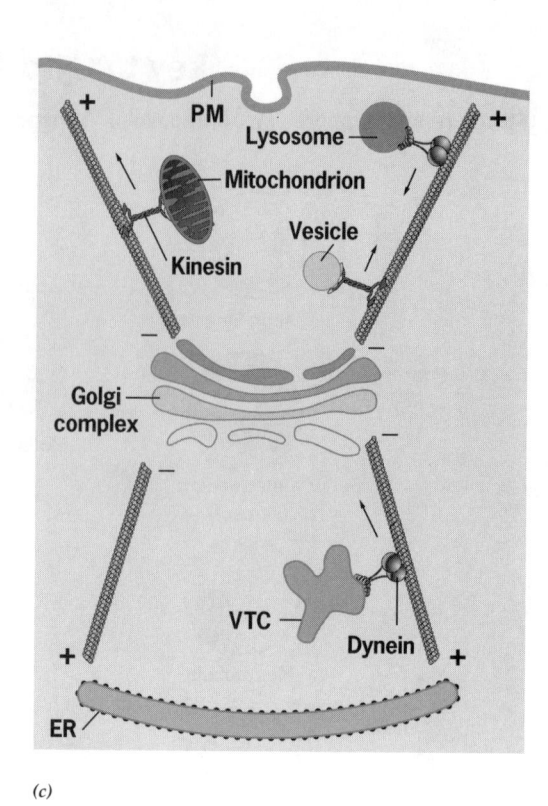

Figure 9.18b,c Cytoplasmic dynein.

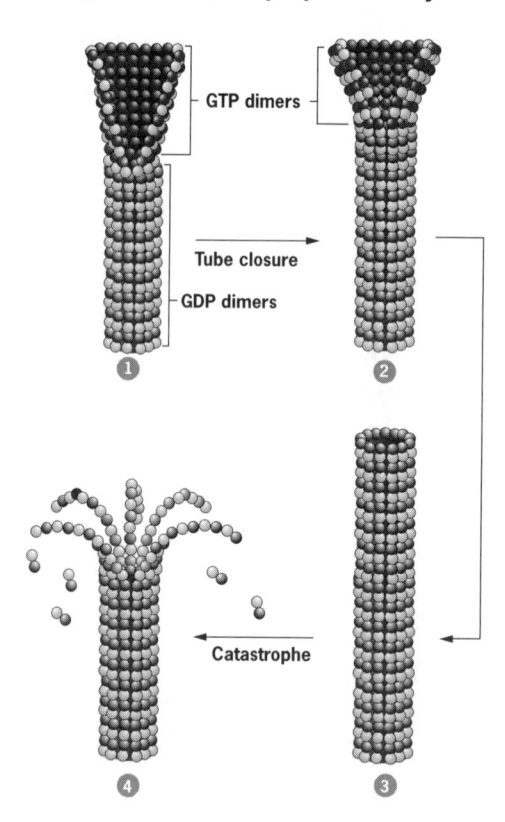

Figure 9.26 The structural cap model of dynamic instability.

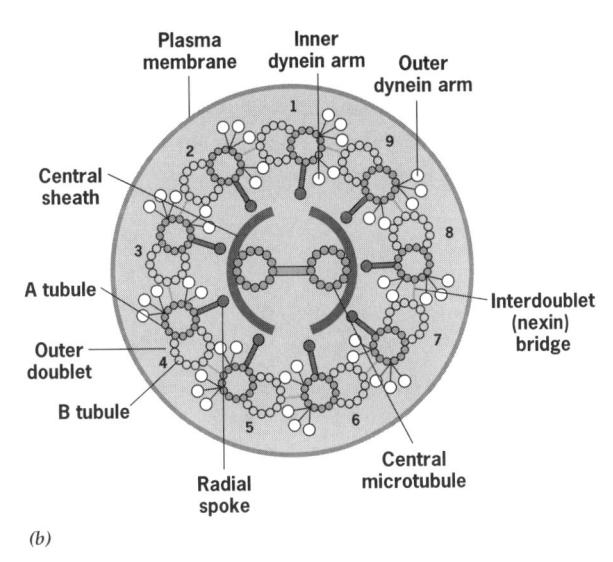

Figure 9.31a The structure of a ciliary or flagellar axoneme.

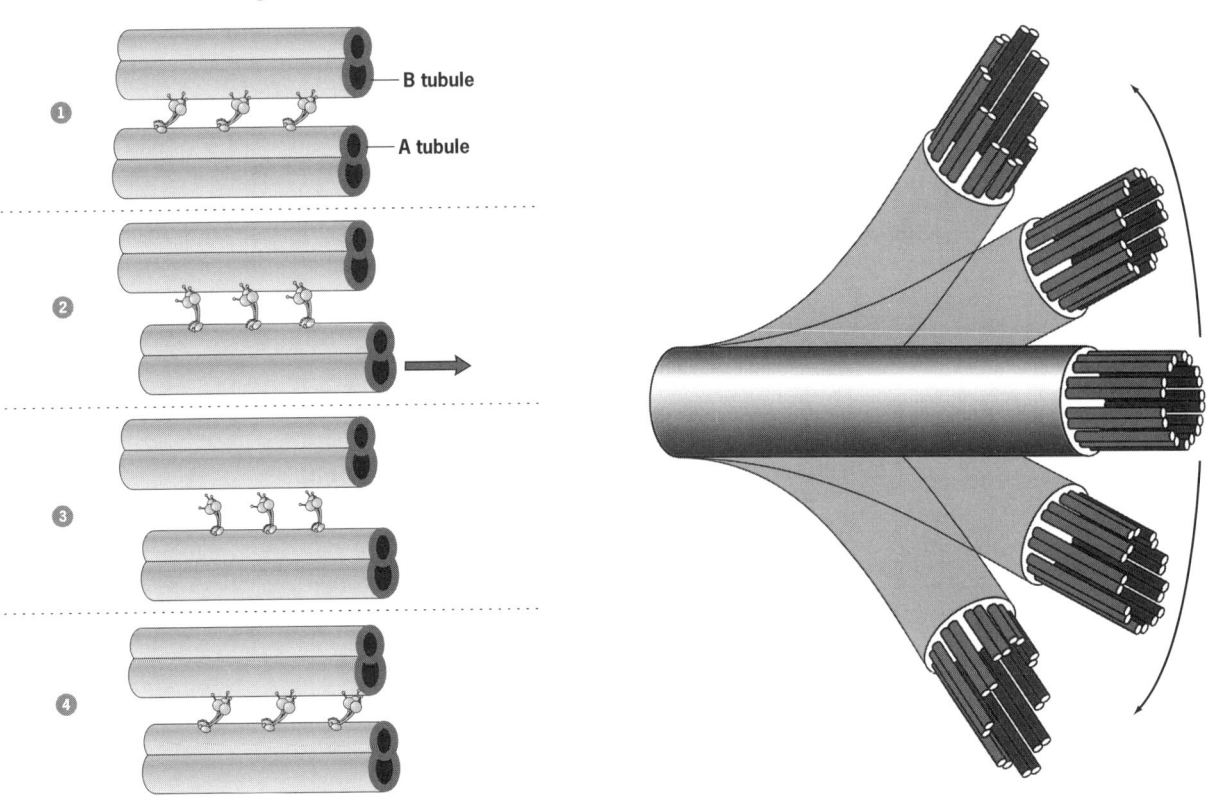

Figure 9.37 The role of dynein arms in generating the force that drives ciliary or flagellar motility.

Figure 9.38a The sliding-microtubule mechanism of ciliary or flagellar motility.

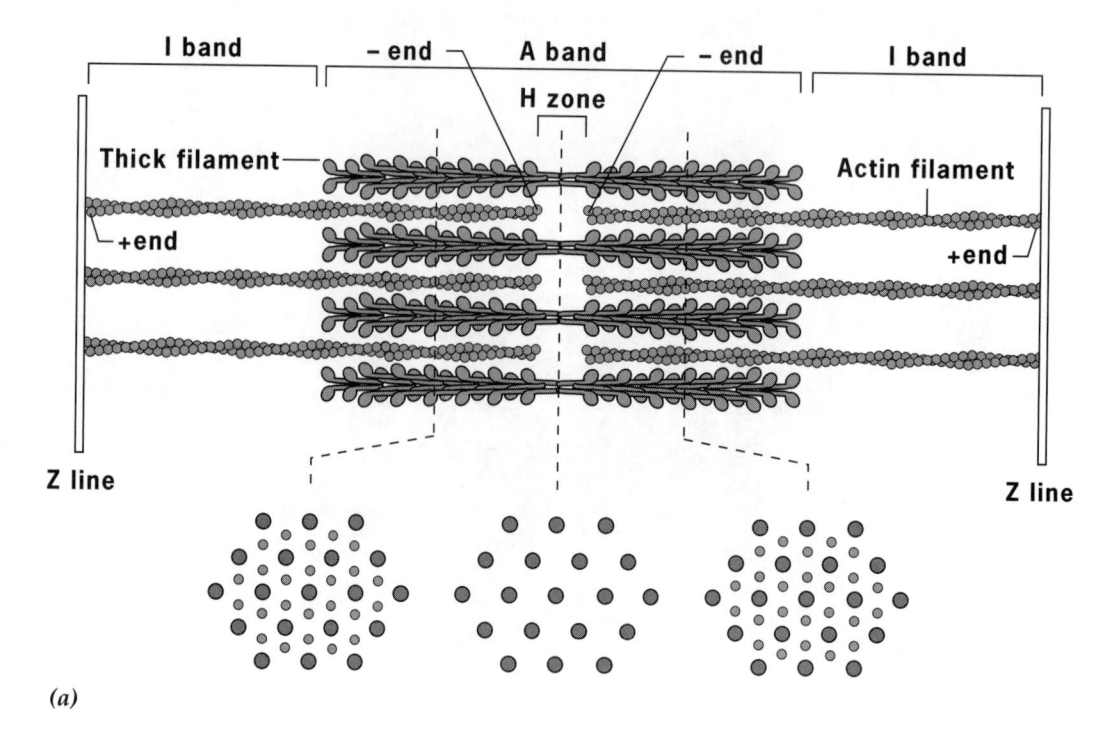

Figure 9.57a The contractile machinery of a sarcomere.

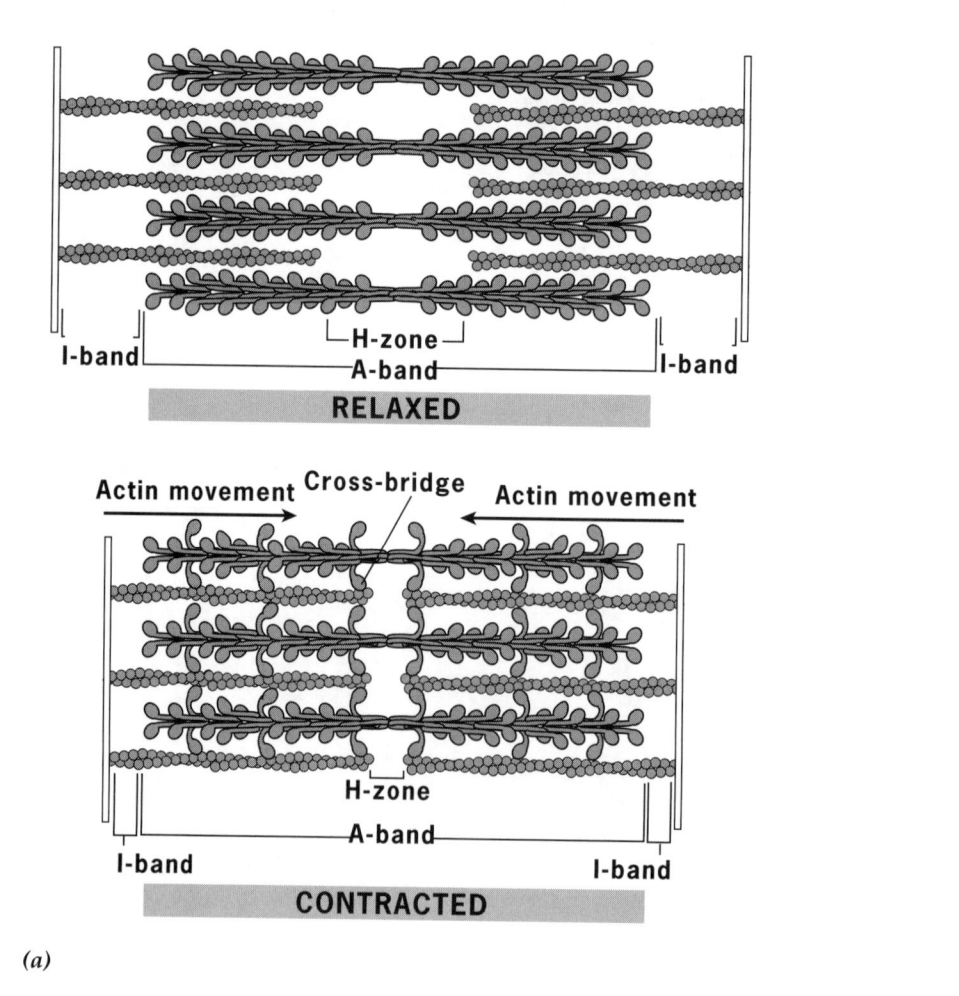

Figure 9.58a The shortening of the sarcomere during muscle contraction.

Copyright © 2002 John Wiley & Sons, Inc.

Figures 9.57a & 9.58a

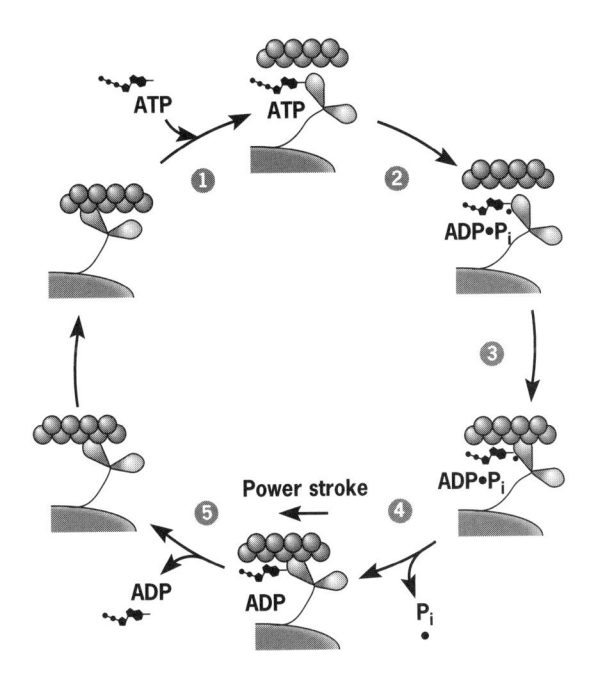

Figure 9.62 A schematic model of the actinomyosin contractile cycle.

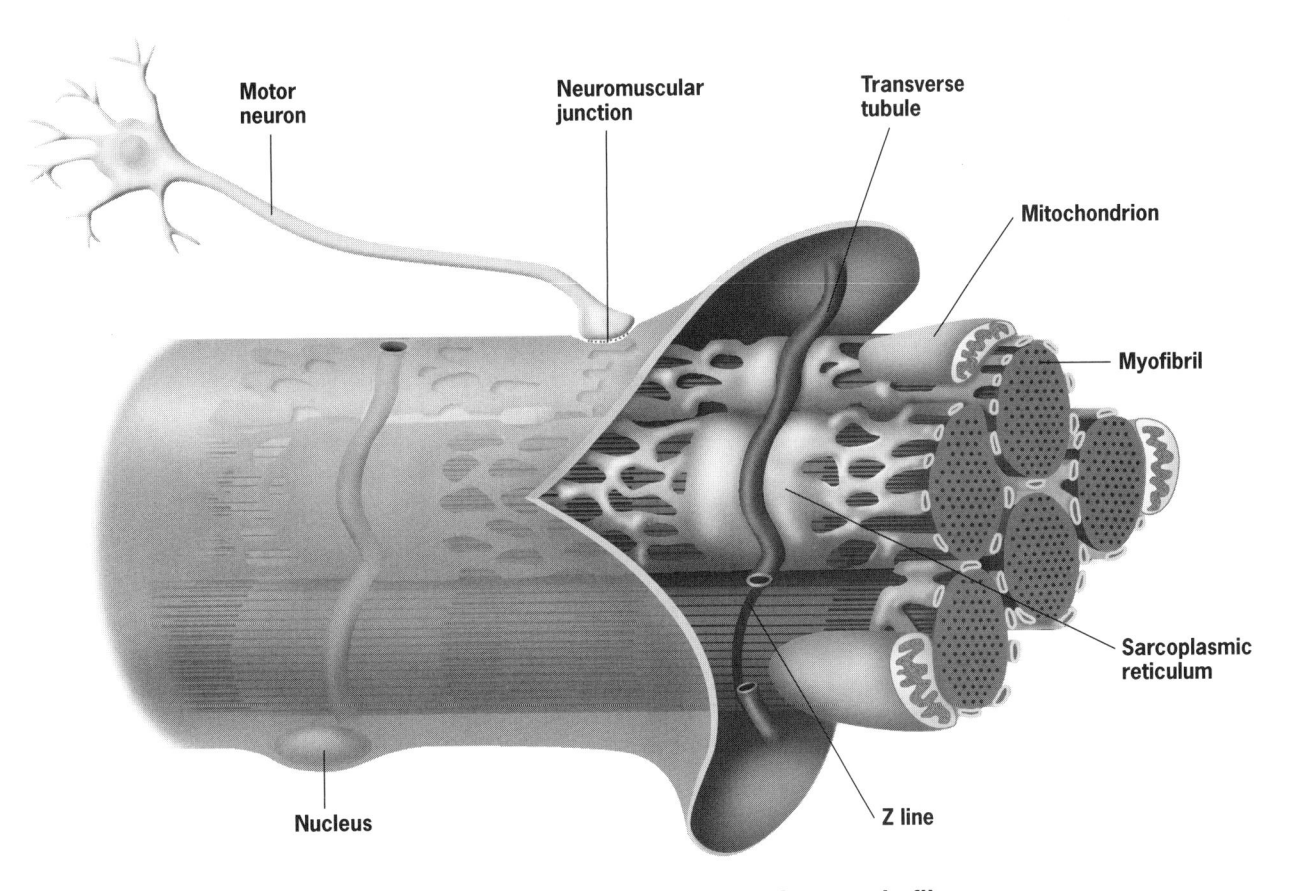

Figure 9.63 The functional anatomy of a muscle fiber.

Figures 9.62 & 9.63

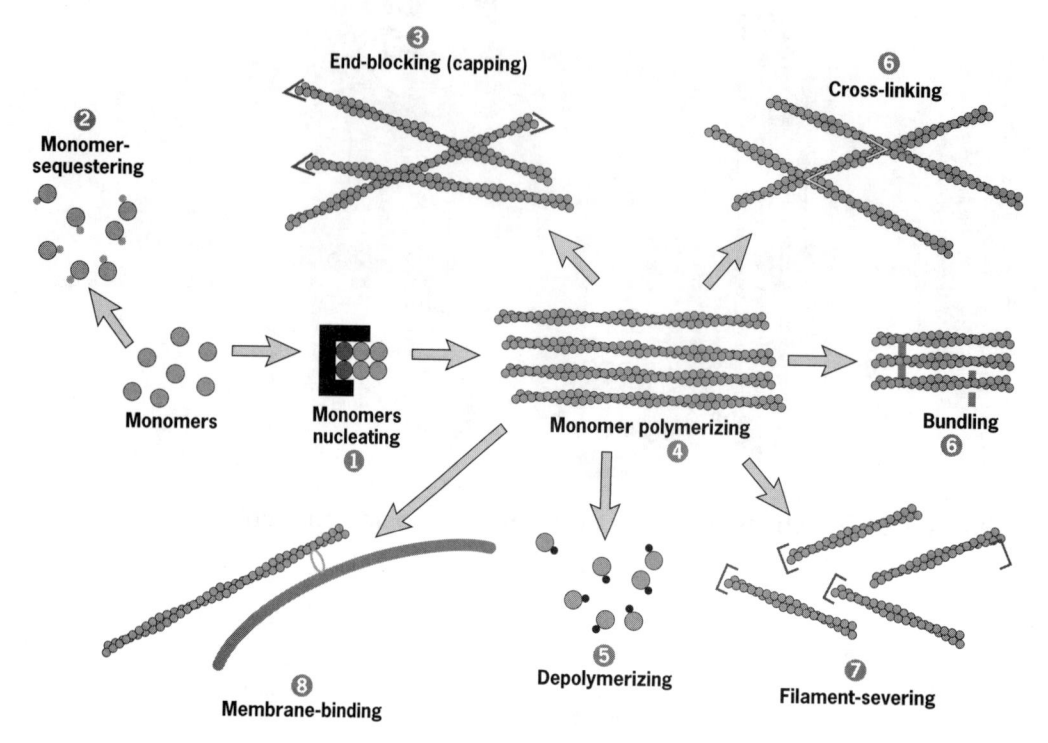

Figure 9.66 The roles of actin-binding proteins.

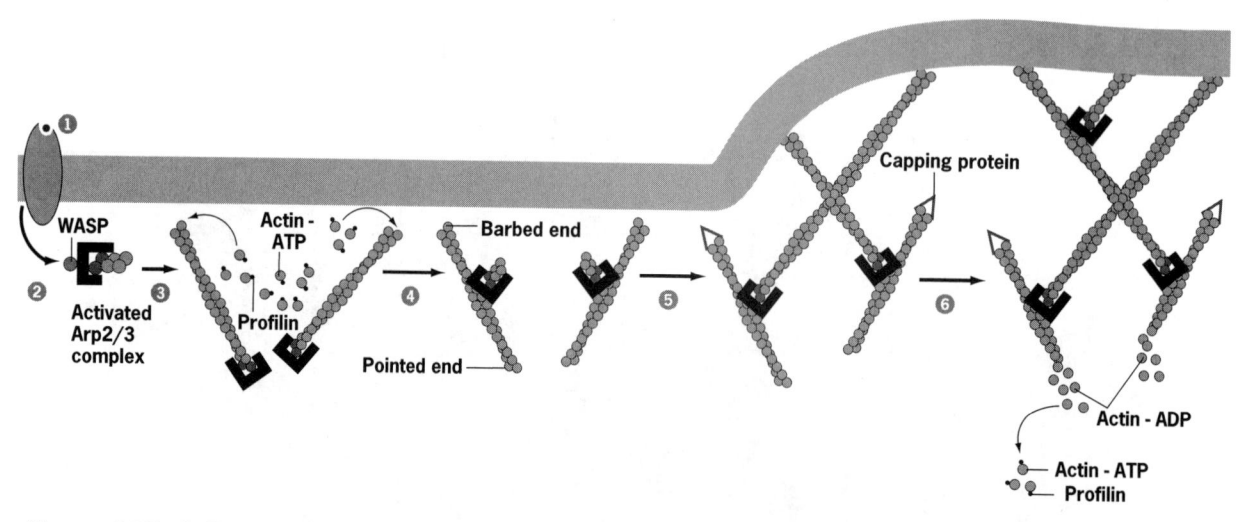

Figure 9.73 A Proposed mechanism for the movement of a nonmuscle cell in a directed manner.

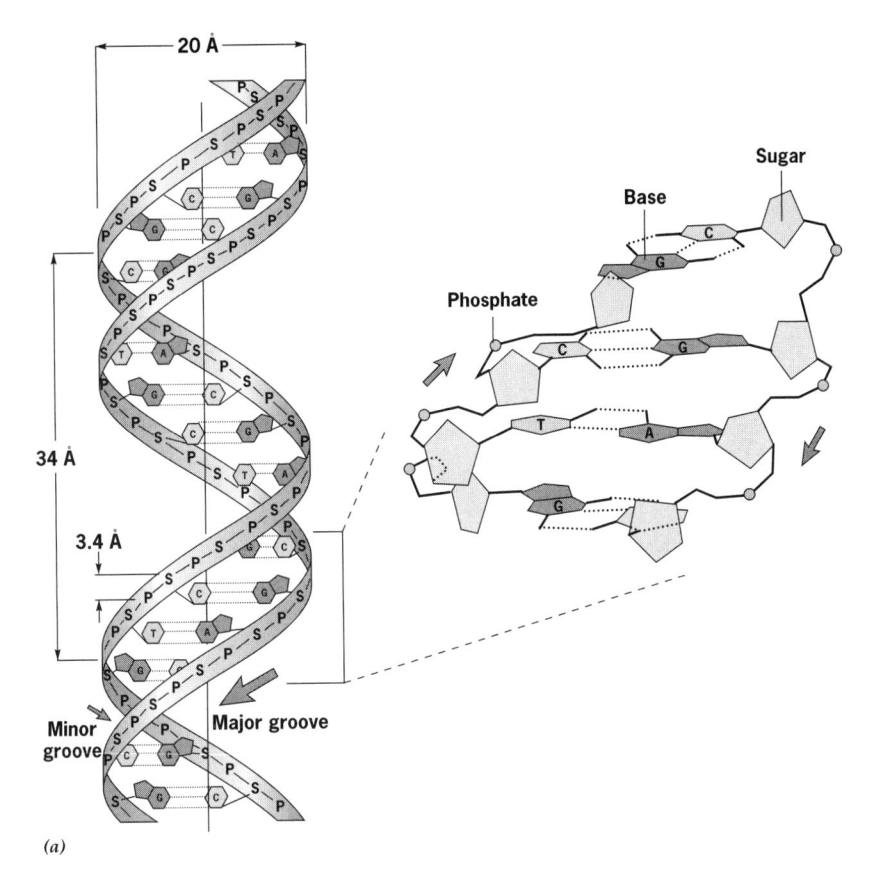

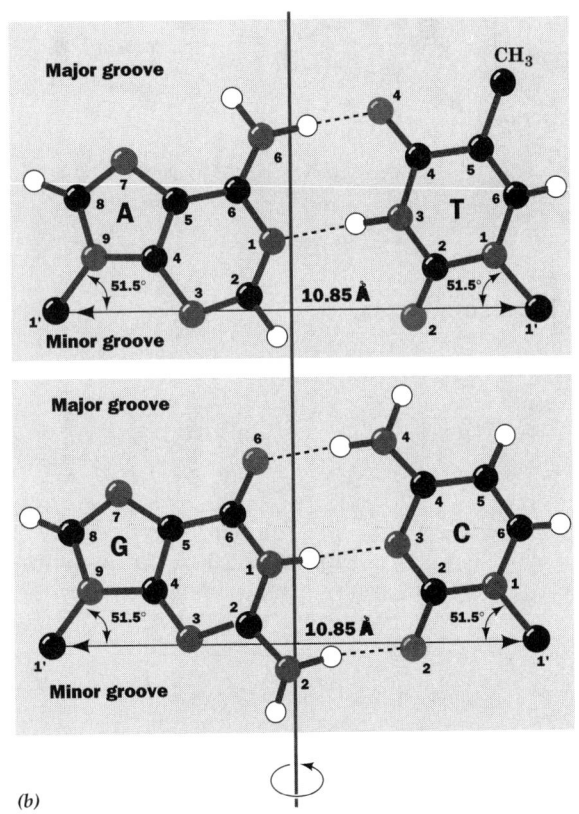

Figure 10.10a,b The double helix.

Figure 10.14b DNA Topoisomerases.

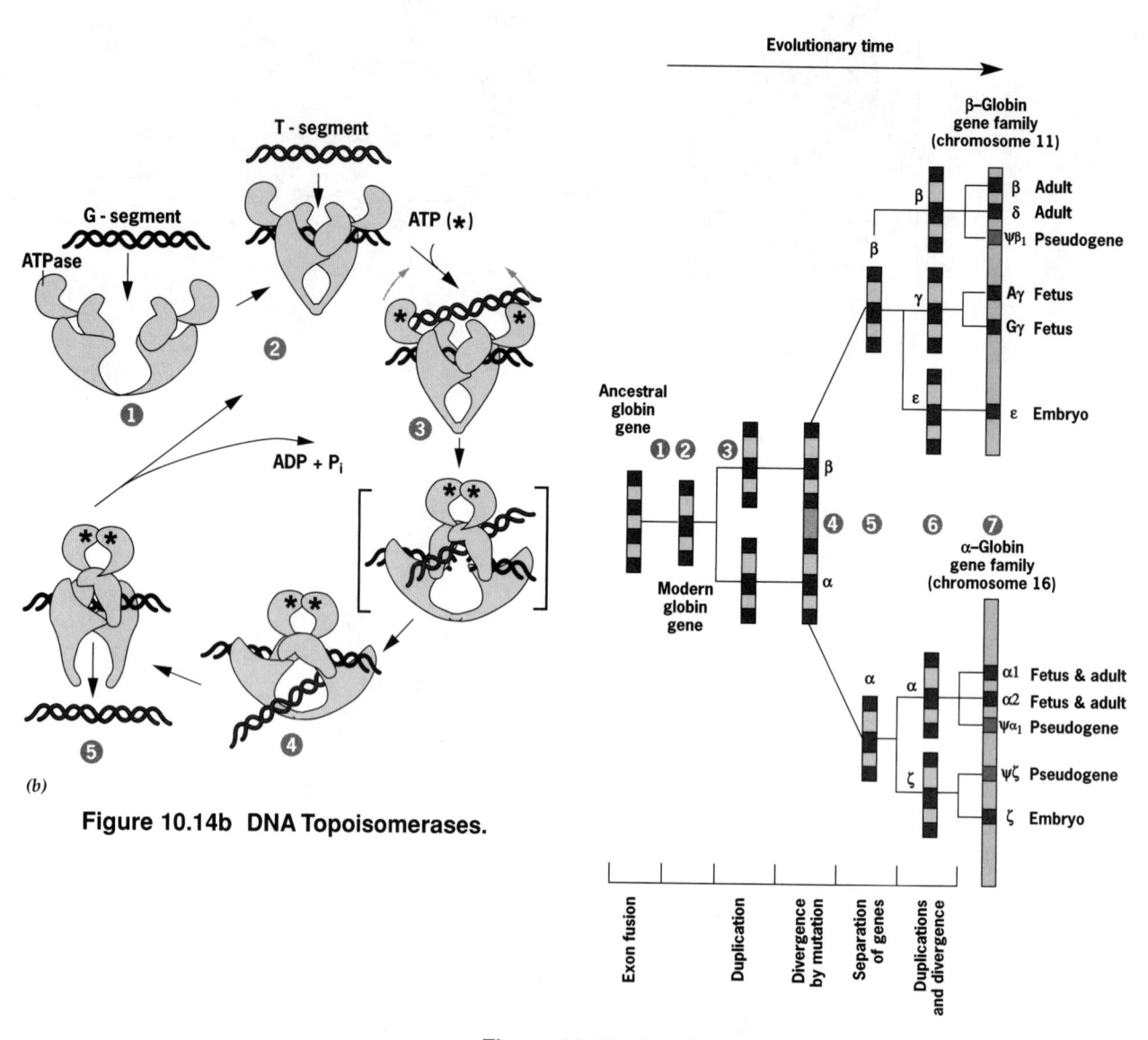

Figure 10.26 A pathway for the evolution of globin genes.

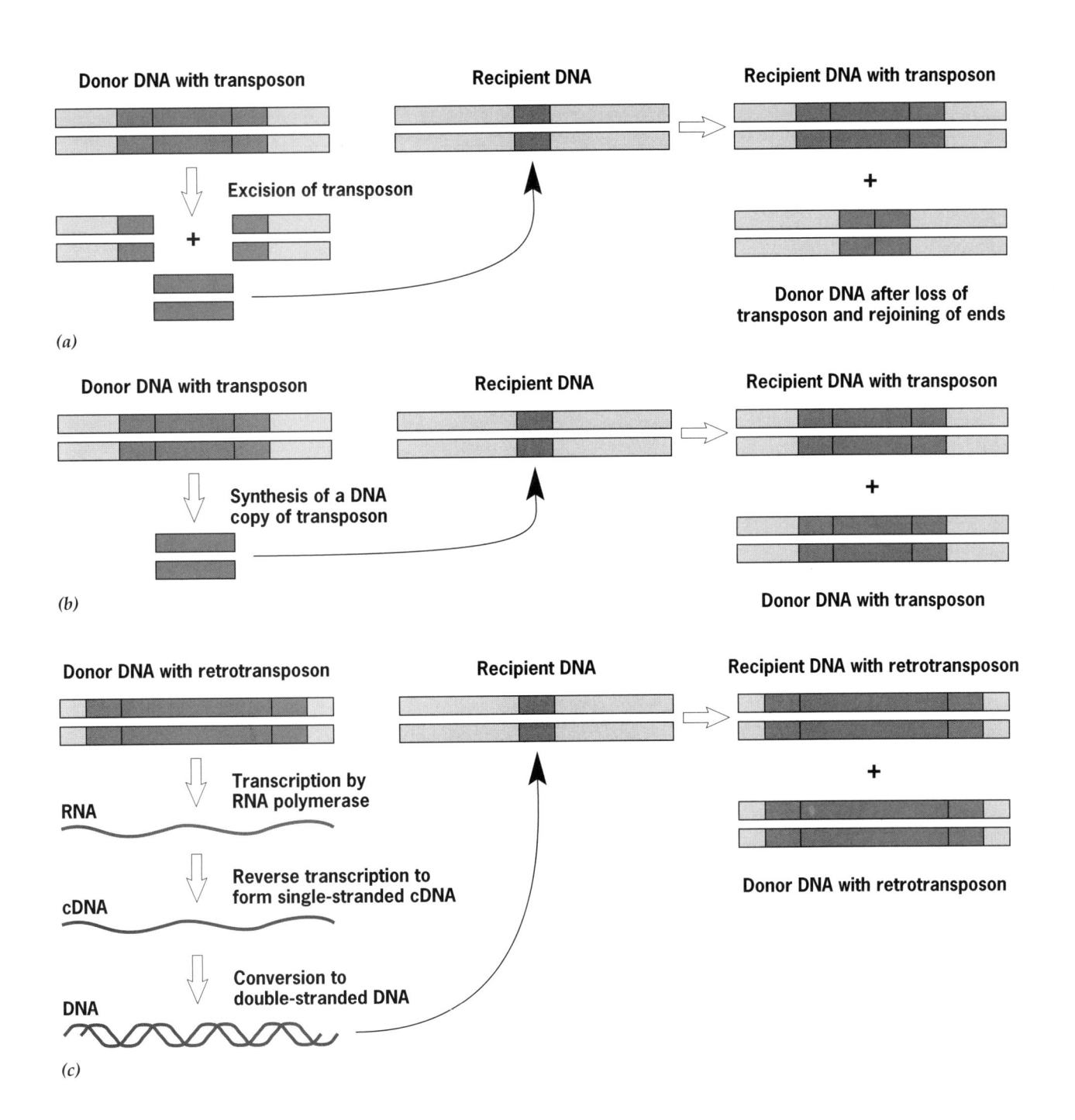

Figure 10.30 Three alternate pathways by which transposable elements move from place to place within the genime.

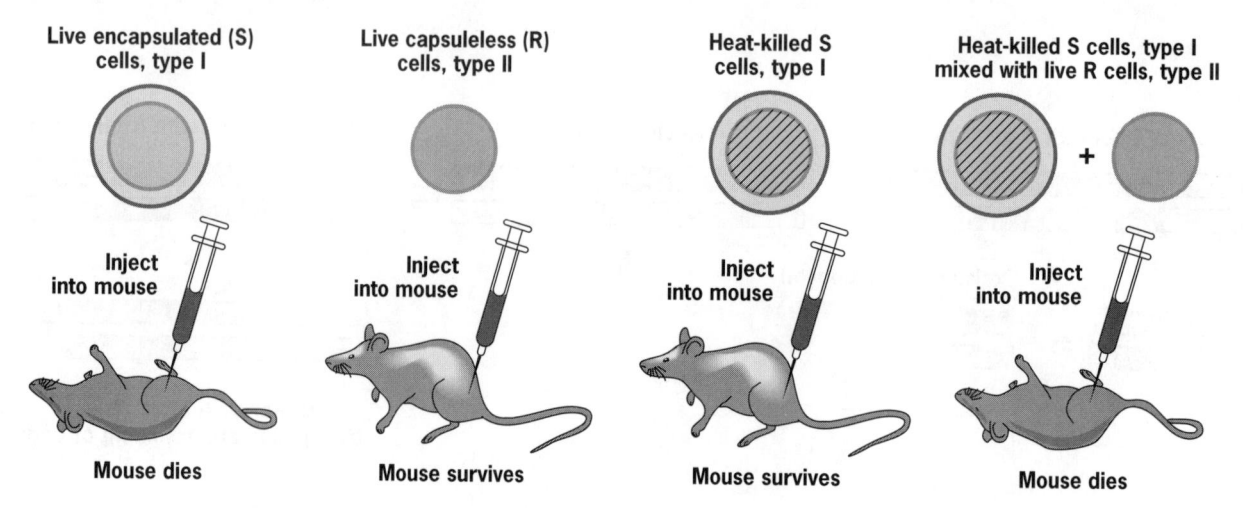

Figure EP2 Outline of the experiment by Griffith of the discovery of bacterial transformation.

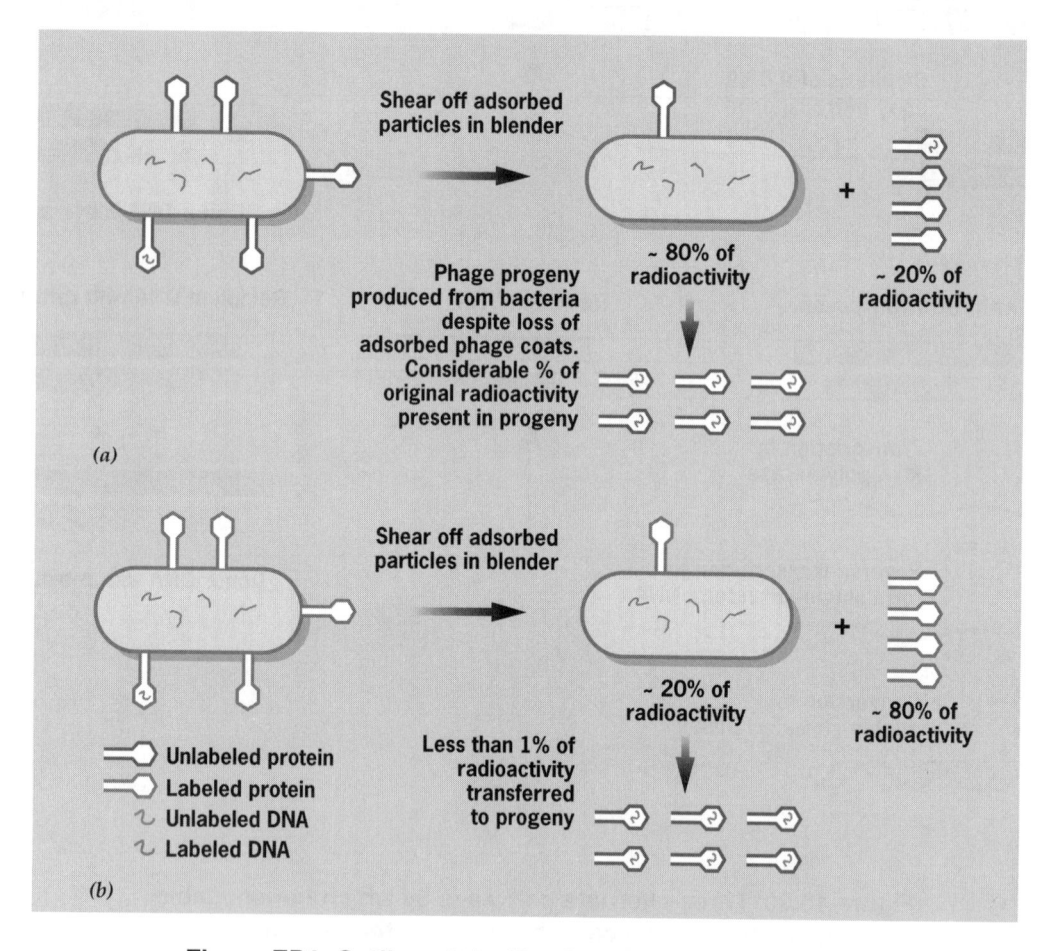

Figure EP4 Outline of the Hershey-Chase experiment.

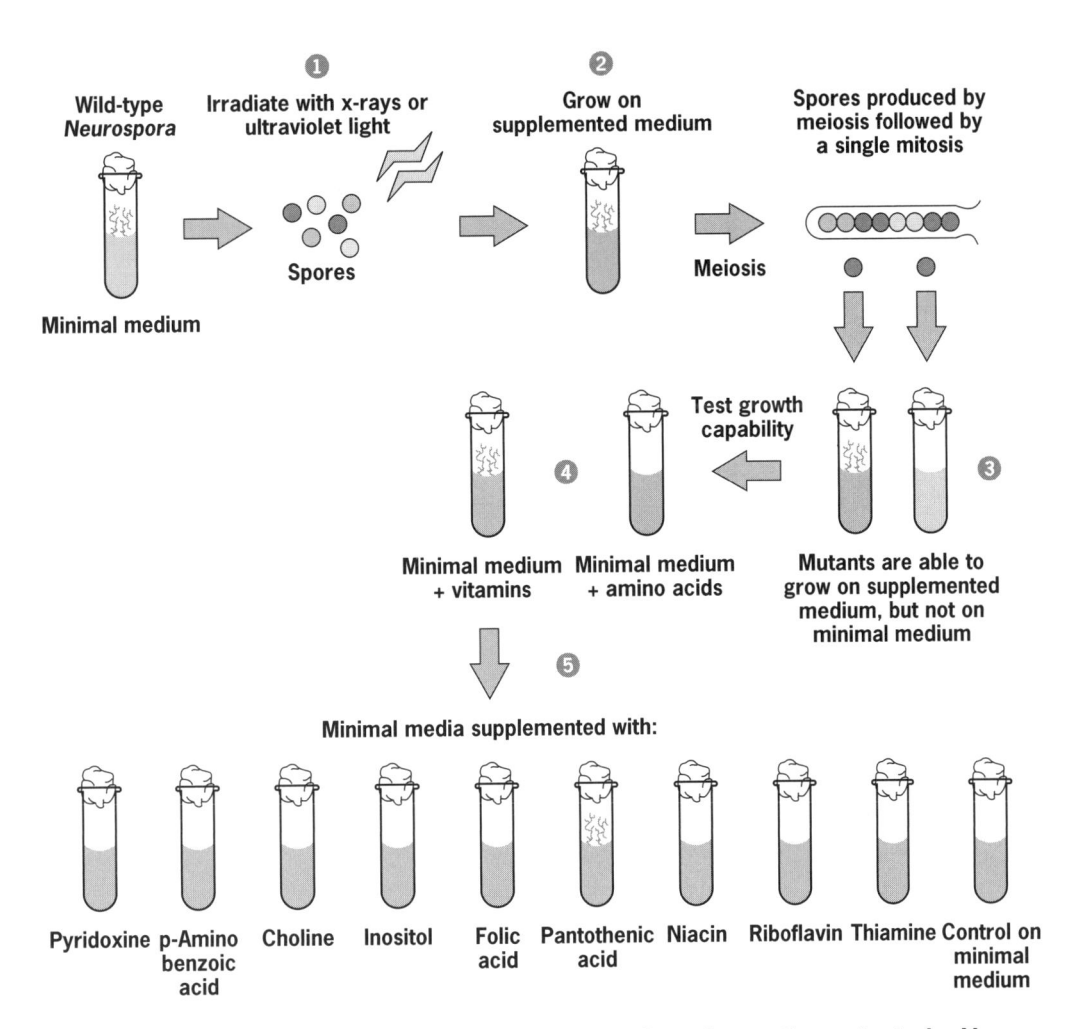

Figure 11.1 The Beadle-Tatum experiment for the isolation of genetic mutants in *Neurospora*.

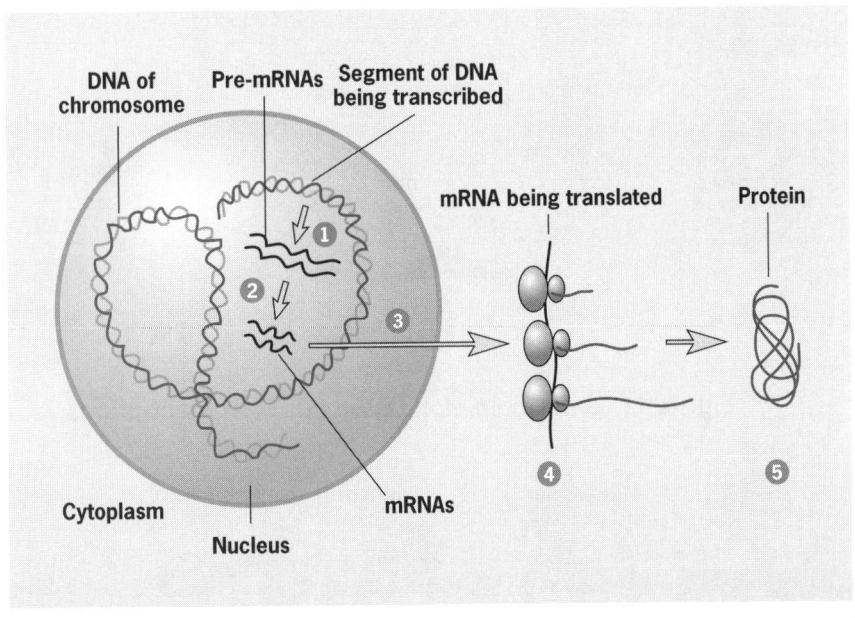

Figure 11.2 An overview of the flow of information in a eukaryotic cell.

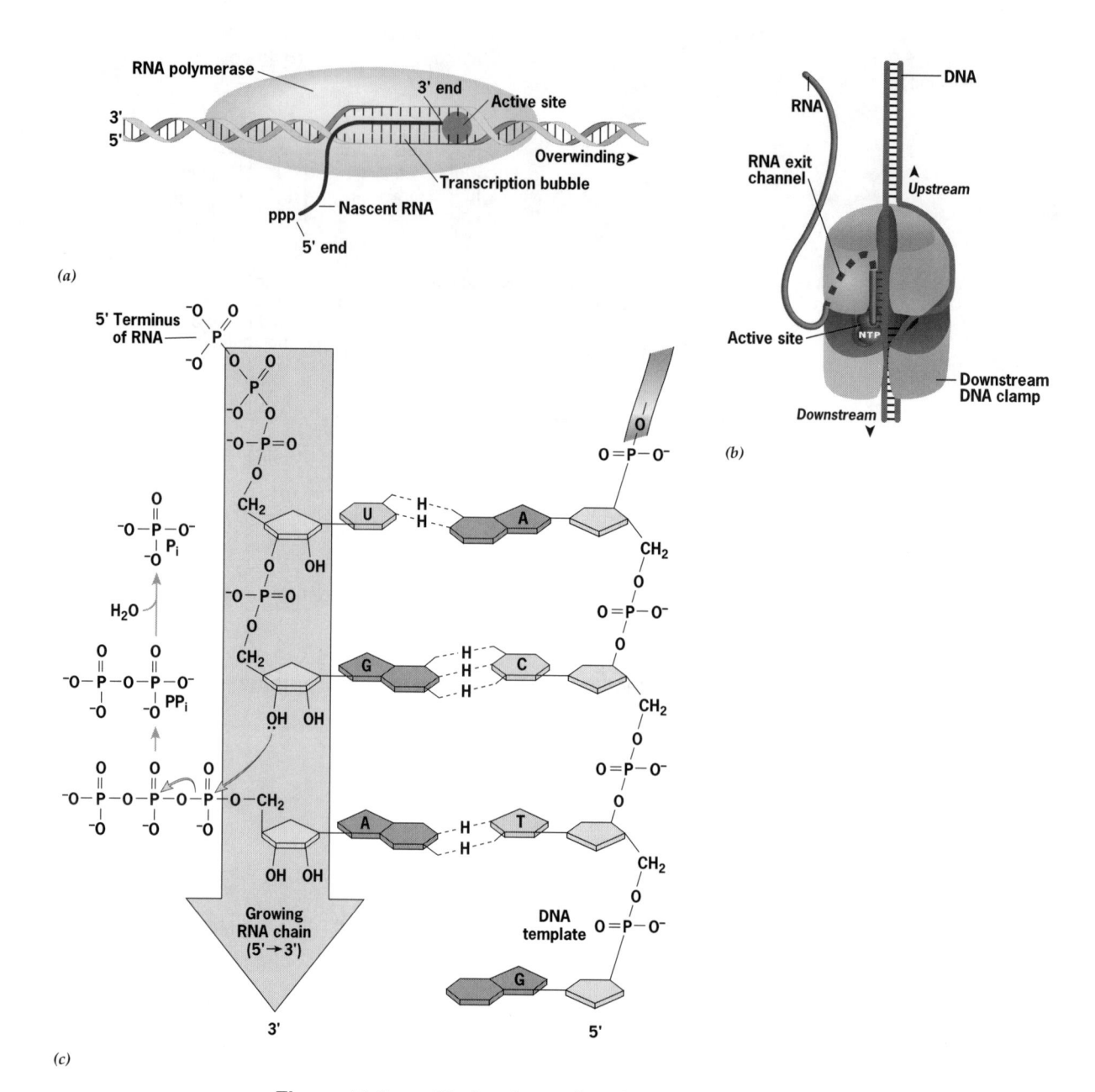

Figure 11.4a-c Chain elongation during transcription.

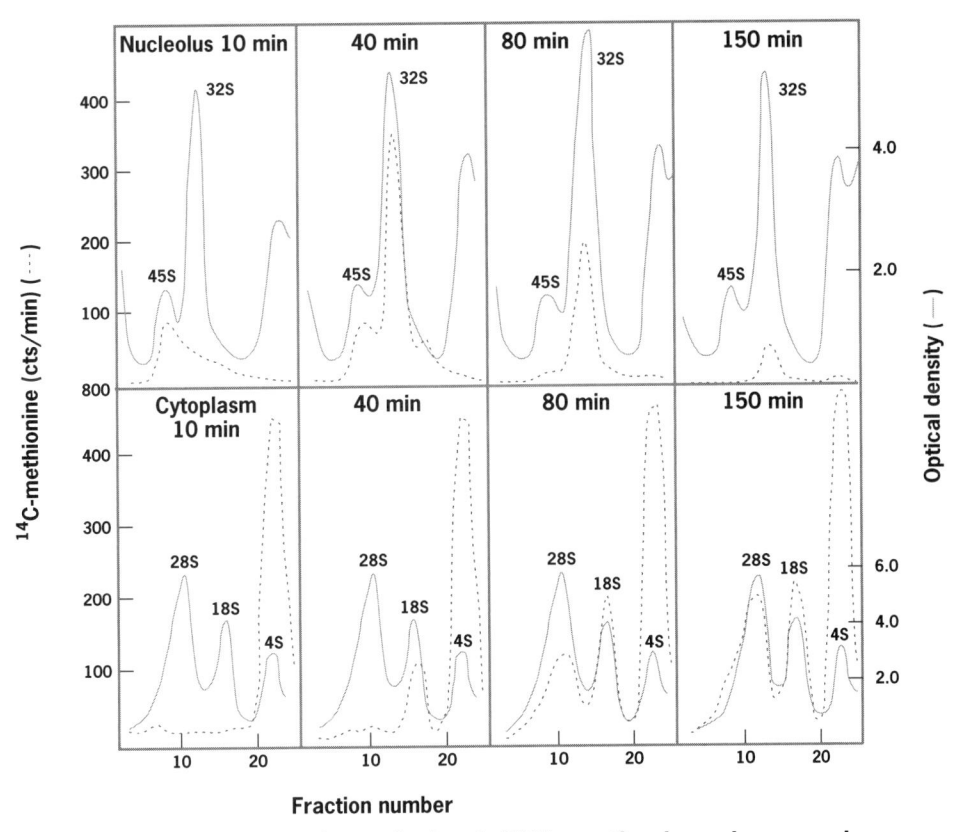

Figure 11.13 Kinetic analysis of rNRA synthesis and processing.

(a)

(b)

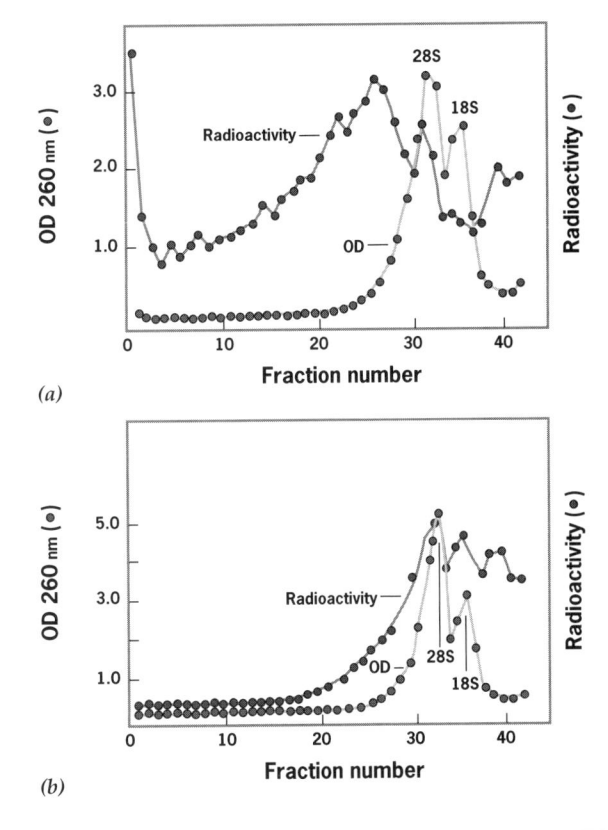

Figure 11.17 The formation of heterogeneous nuclear RNA (hnRNA) and its conversion into smaller mRNAs.

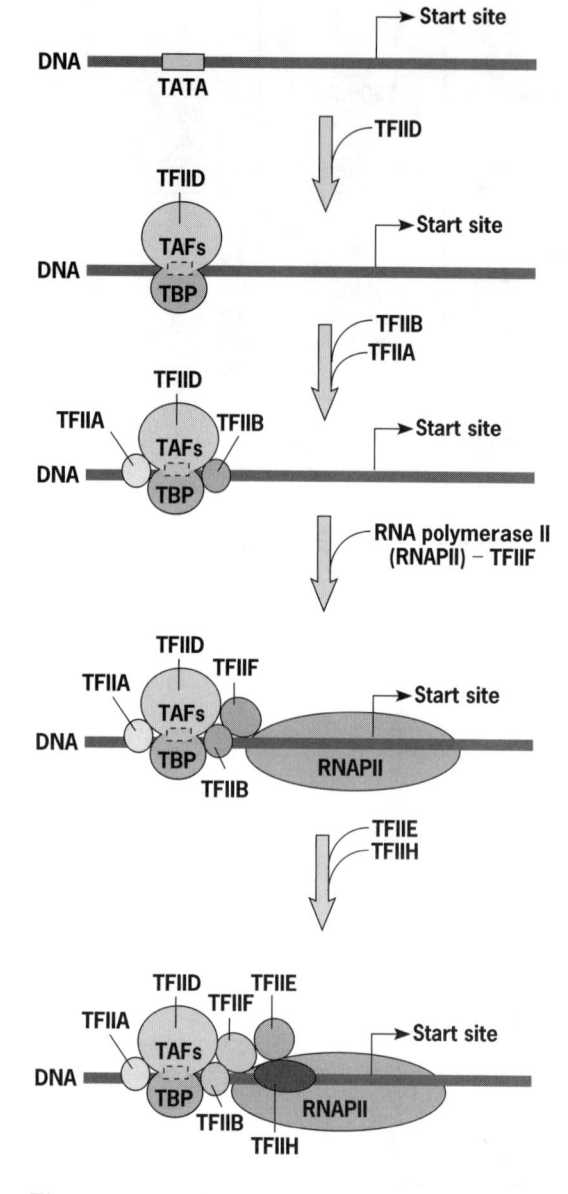

Figure 11.19 A model of the steps in the assembly of the preinitiation complex for RNA polymerase II at the TATA box of a eukaryotic promoter.

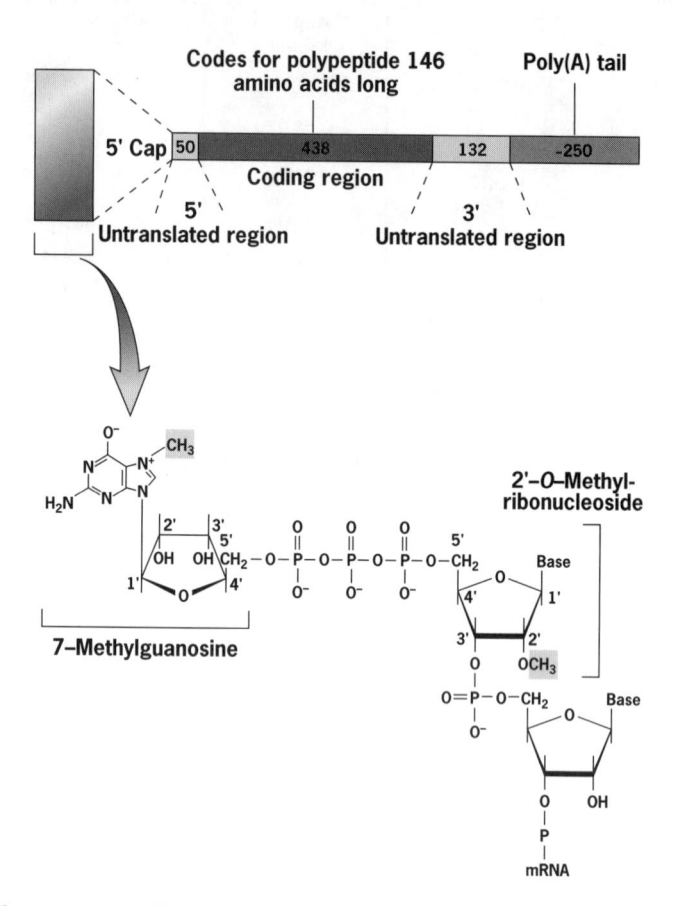

Figure 11.21 Structure of the human β-globin mRNA.

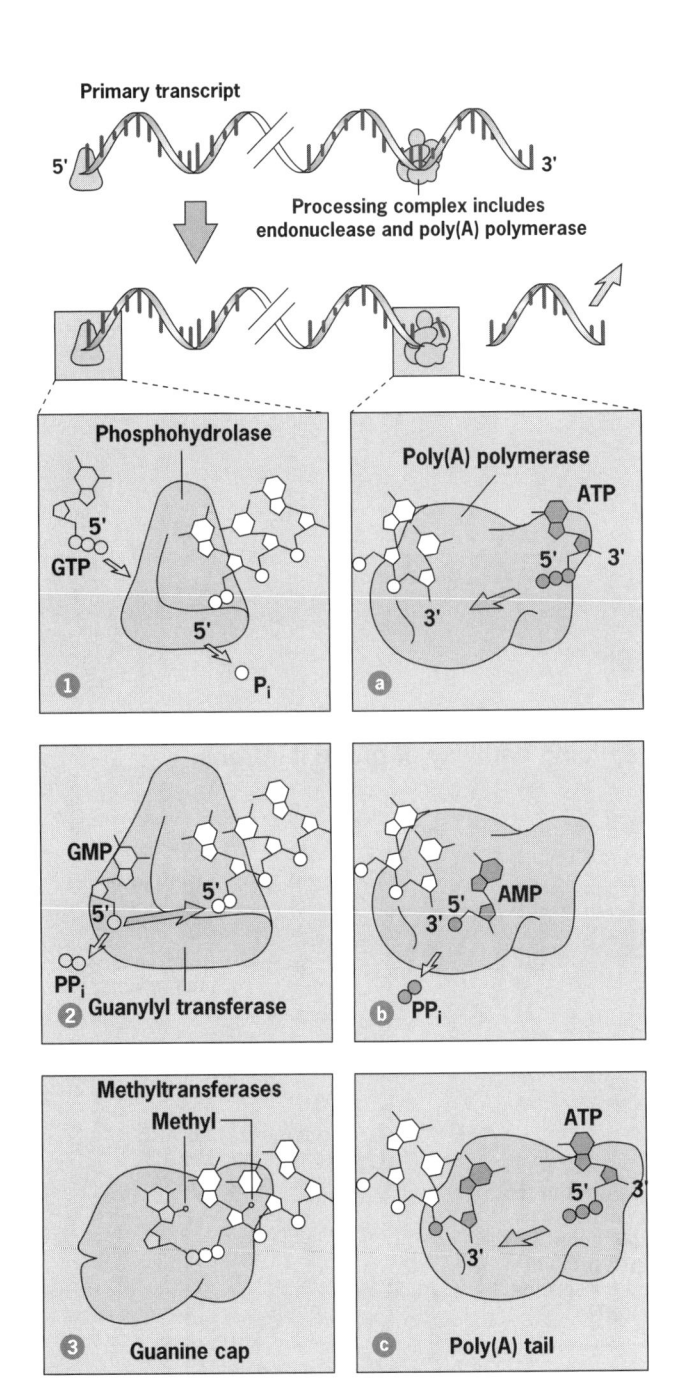

Figure 11.28 Steps in the addition of a 5´ methylguanosine cap and 3´ poly(A) tail to a pre-mRNA.

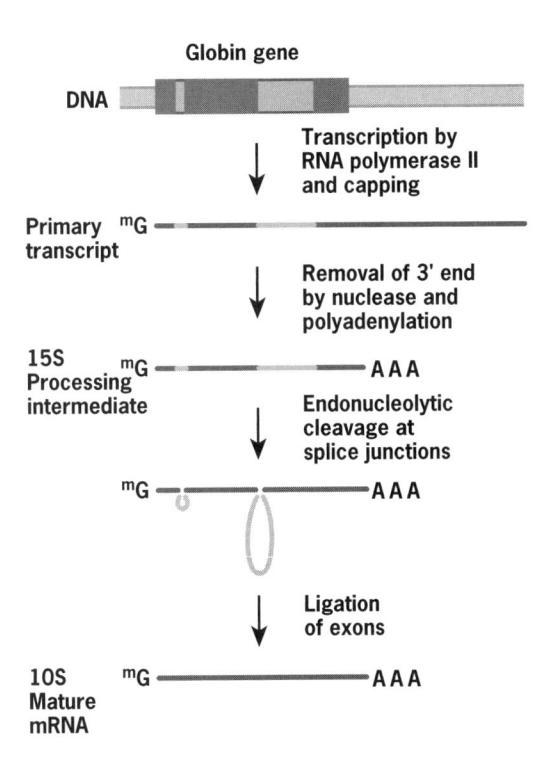

Figure 11.29 Overview of the steps during the processing of the globin mRNA.

Figures 11.28 & 11.29

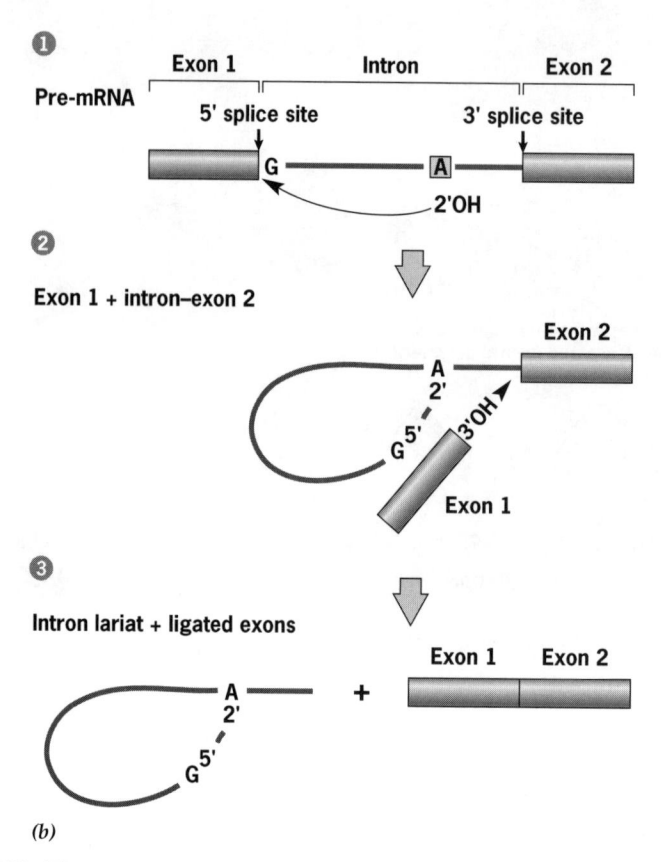

Figure 11.32b The structure and self-splicing pathway of group II introns.

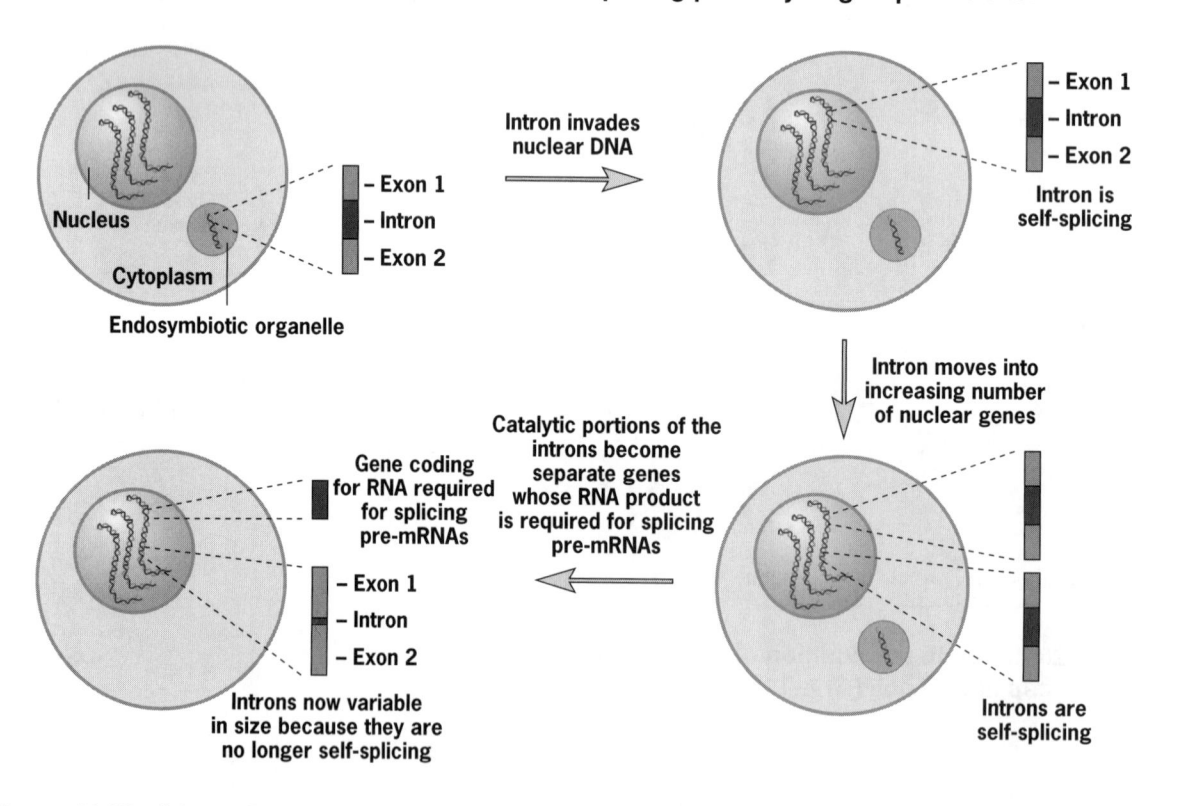

Figure 11.39 A hypothesis that may explain the evolutionary origin of introns in eukaryotic DNA.

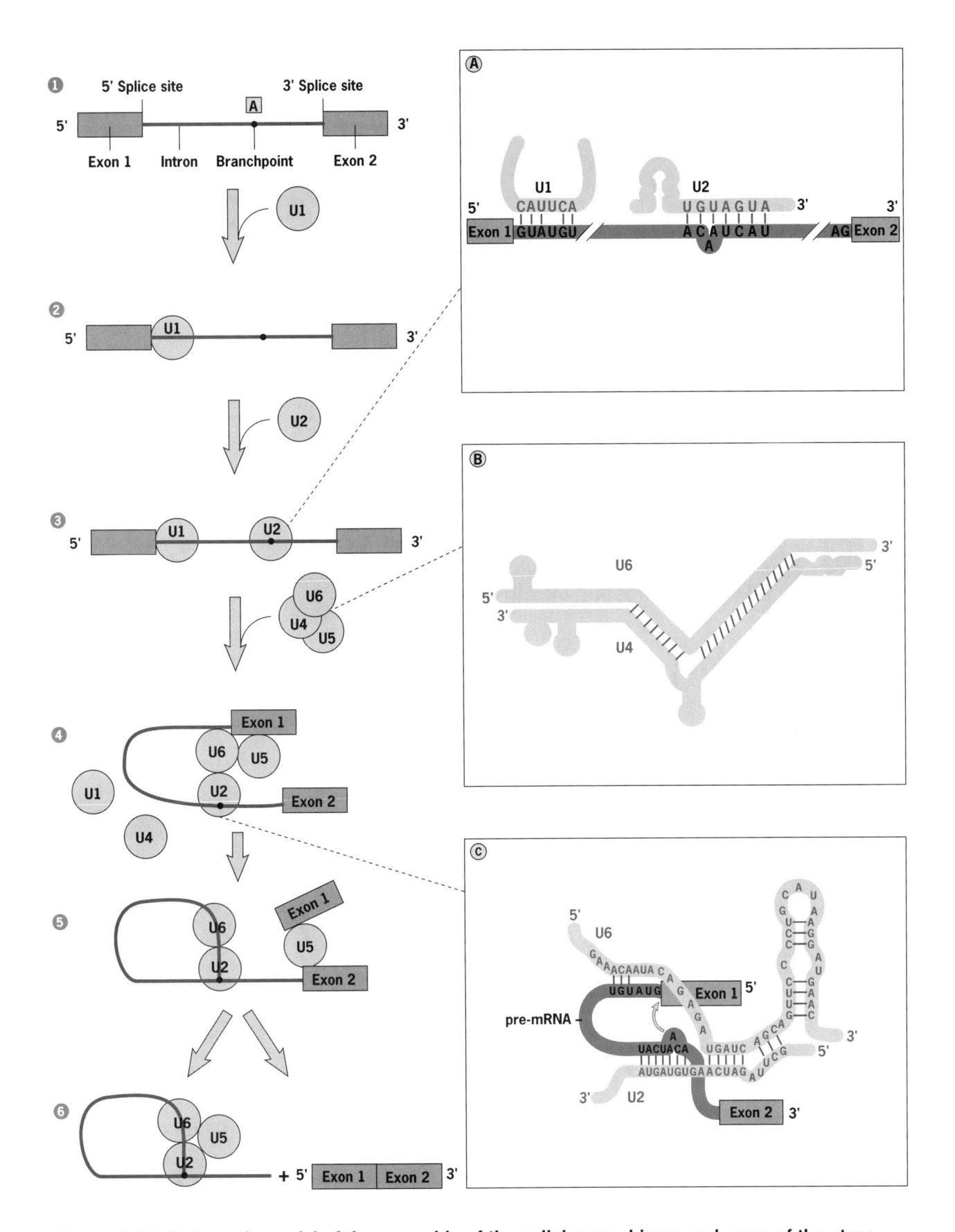

Figure 11.35 Schematic model of the assembly of the splicing machinery and some of the steps that occur during splicing.

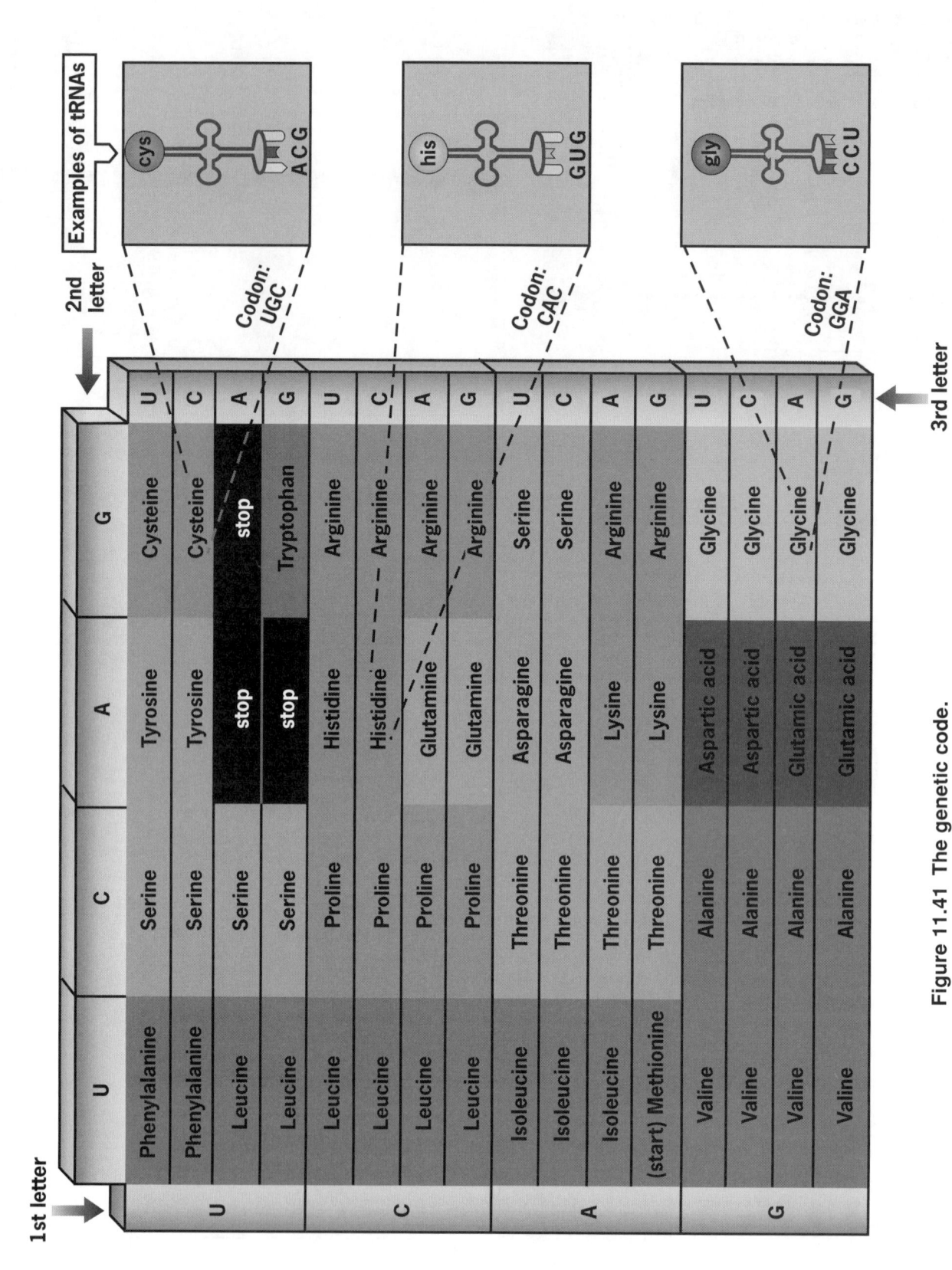

Figure 11.41 The genetic code.

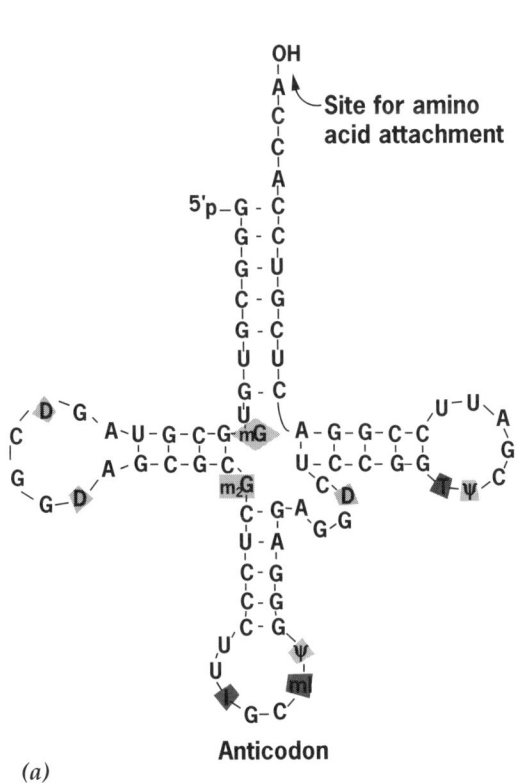

(a) Figure 11.42a Three-dimensional structure of transfer RNAs.

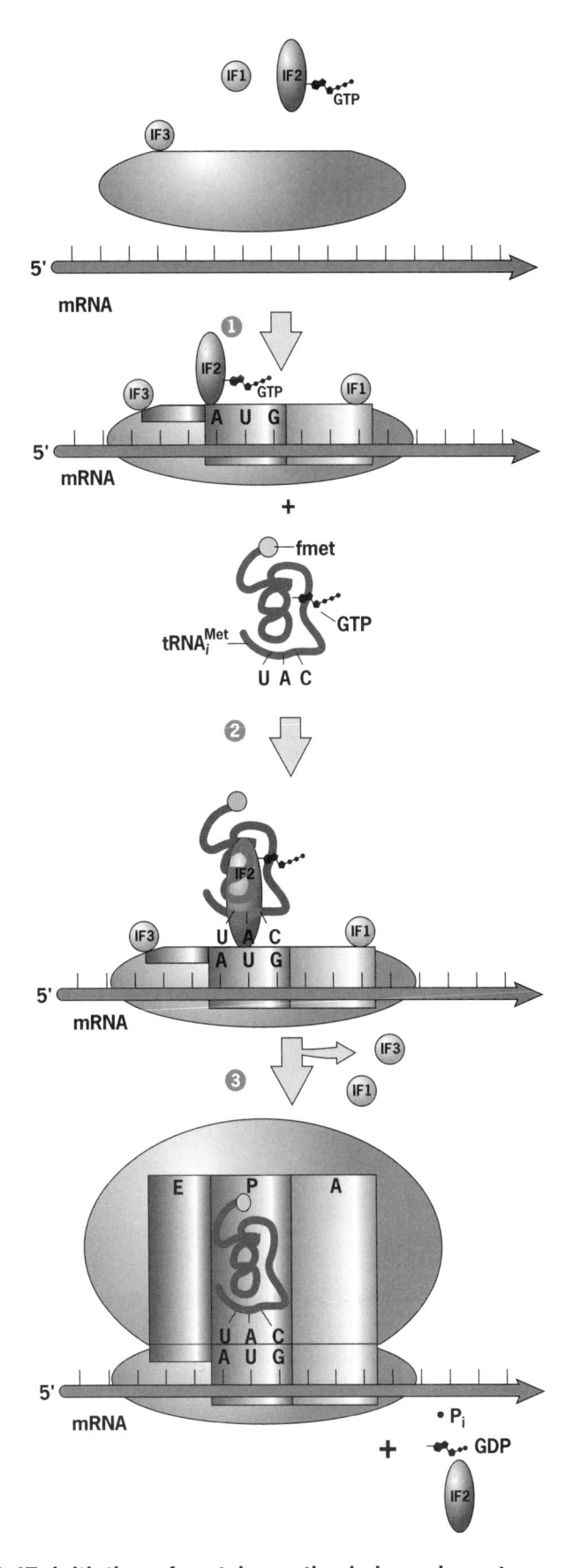

Figure 11.47 Initiation of protein synthesis in prokaryotes.

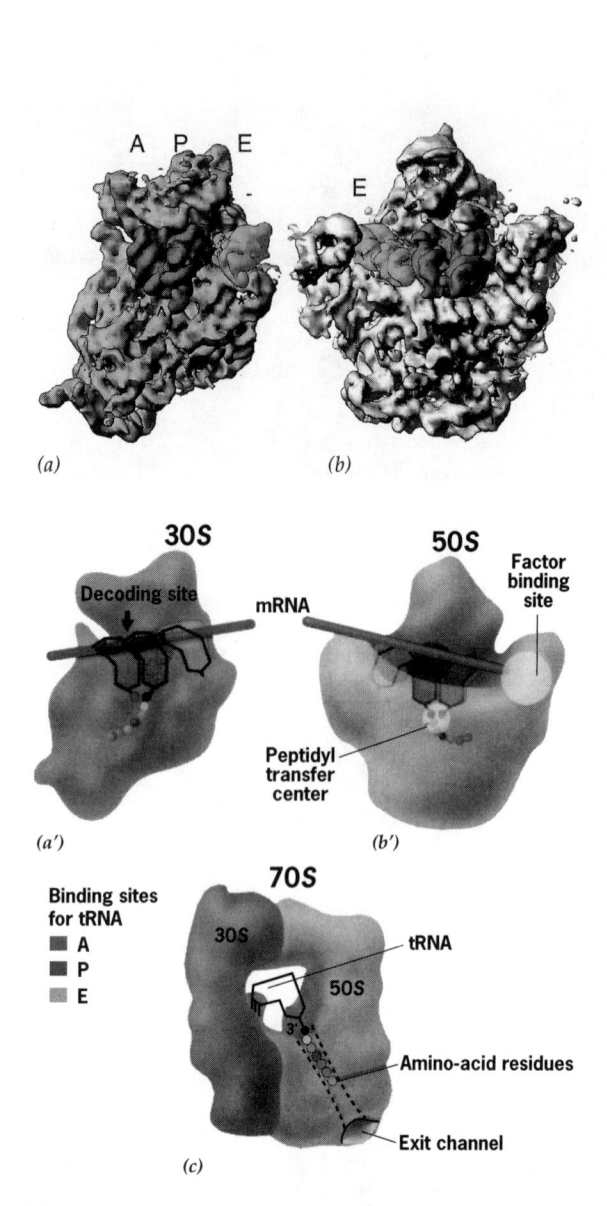

Figure 11.49 Model of the bacterial ribosome based on X-ray crystallographic data, showing tRNAs bound to the A, P, and E sites of the two ribosomal subunits.

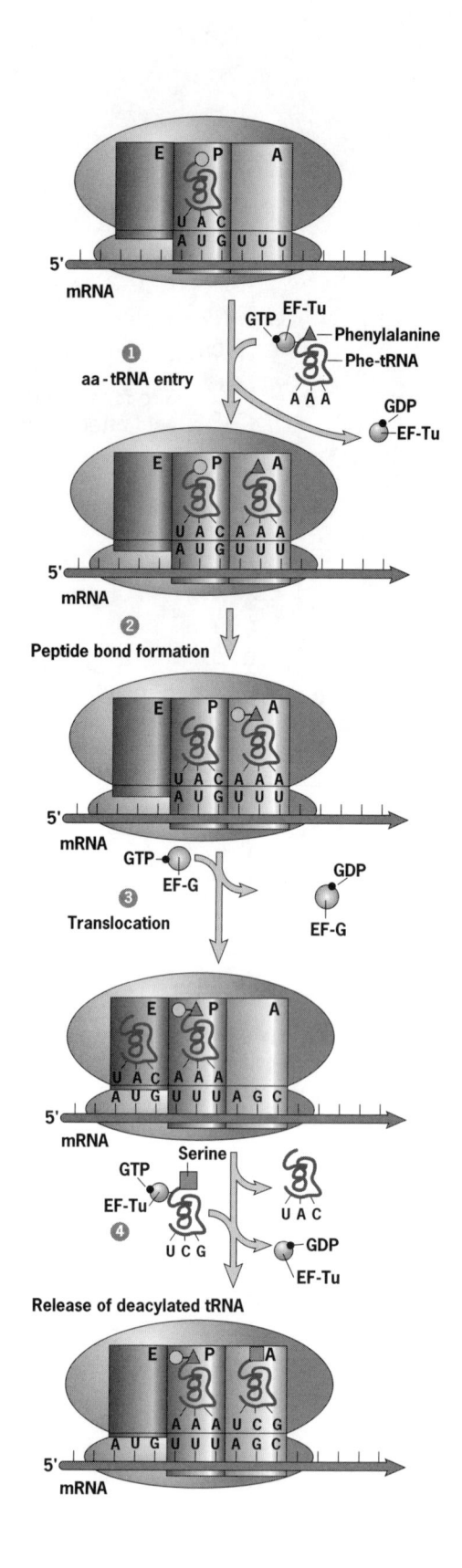

Figure 11.50a Steps in the elongation of the nascent polypeptide during translation in prokaryotes.

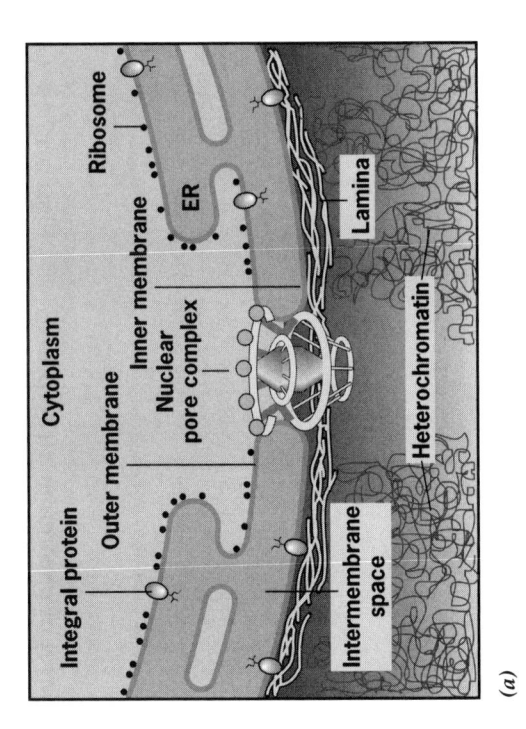

Figure 12.2a The nuclear envelope.

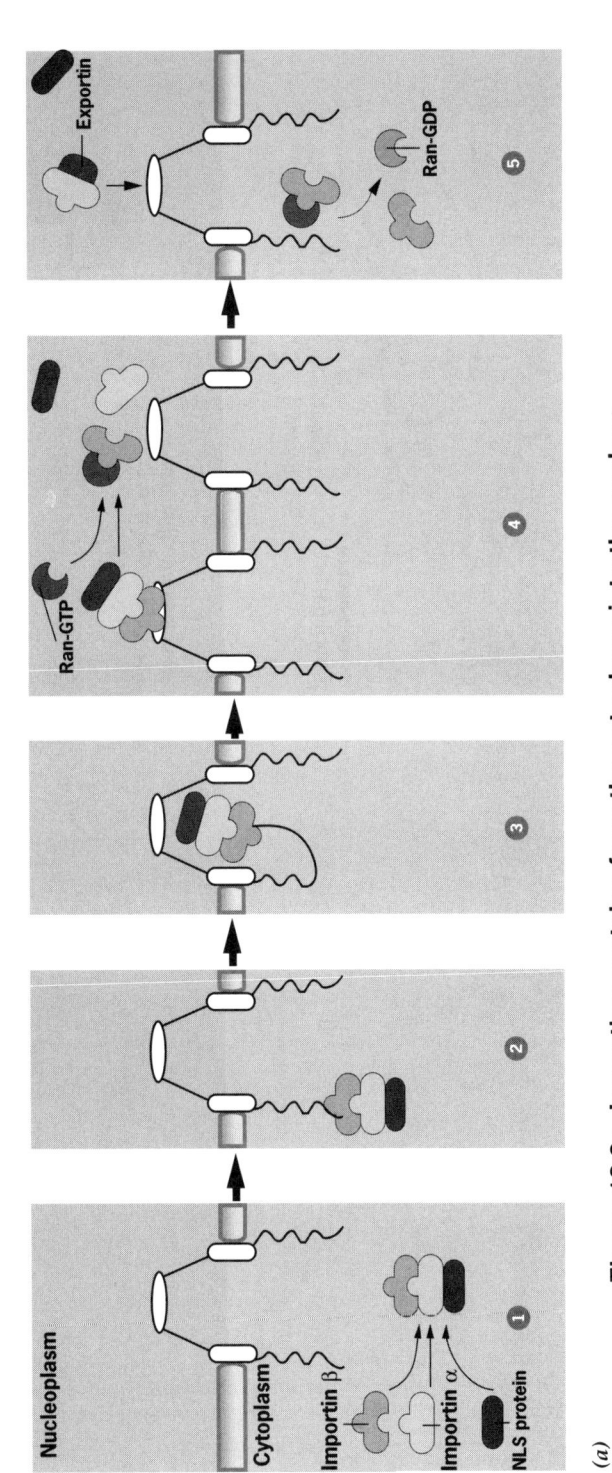

Figure 12.8a Importing proteins from the cytoplasm into the nucleus.

Figures 12.2a & 12.8a

Figure 12.14 Levels of organization of chromatin.

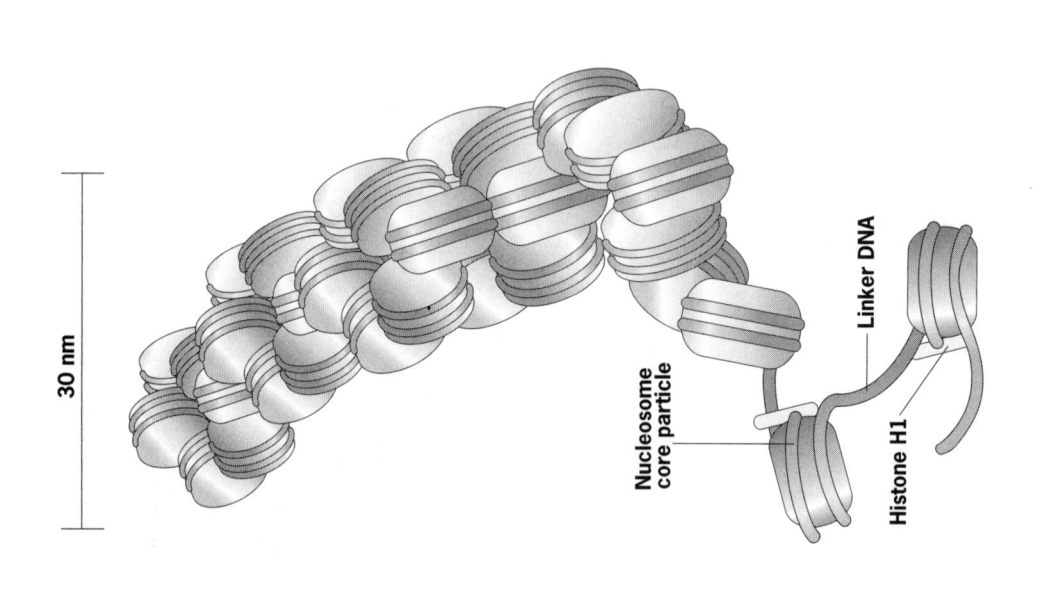

Figure 12.12b The 30-nm fiber: a higher level of chromatin structure.

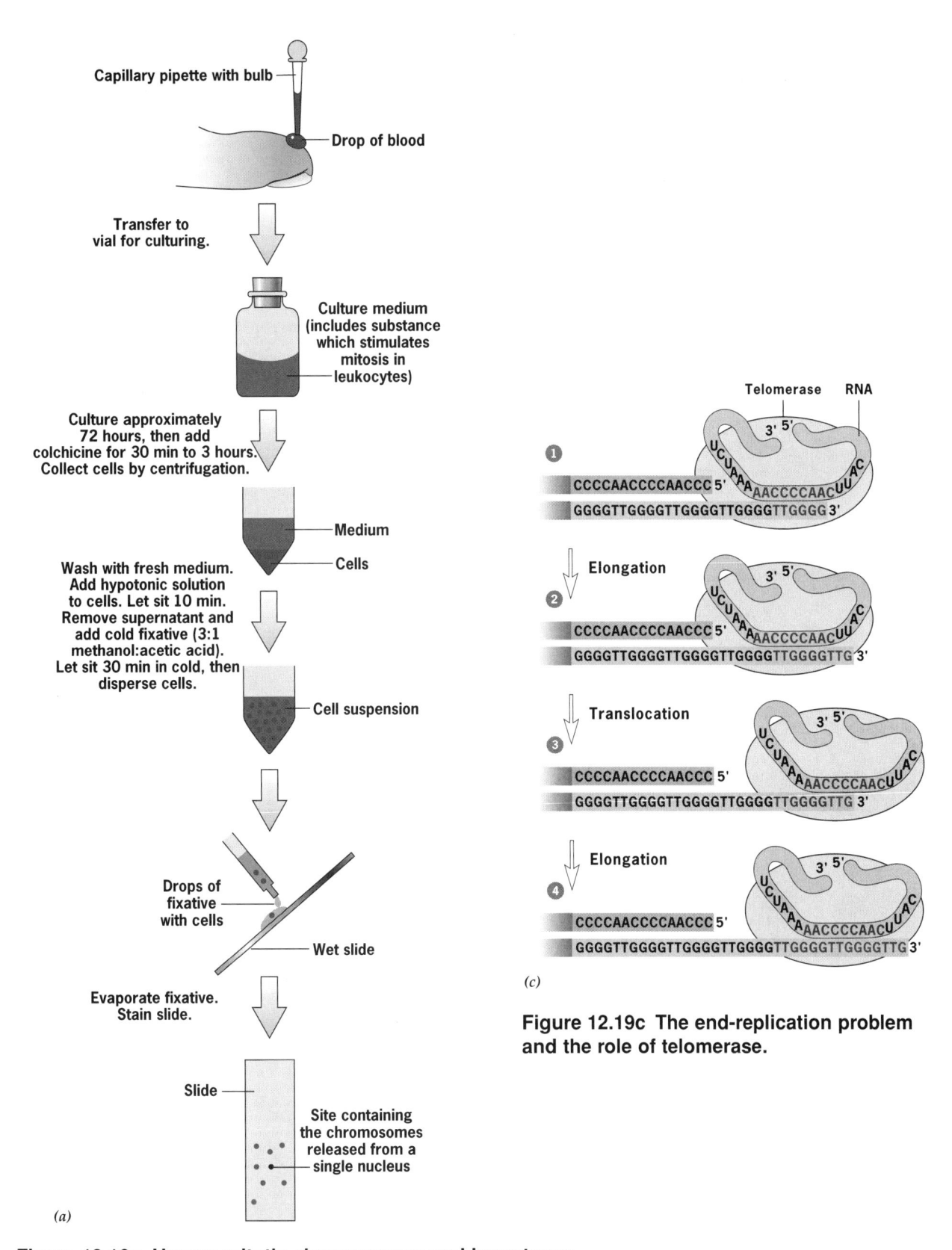

Figure 12.16a Human mitotic chromosomes and karyotypes.

Figure 12.19c The end-replication problem and the role of telomerase.

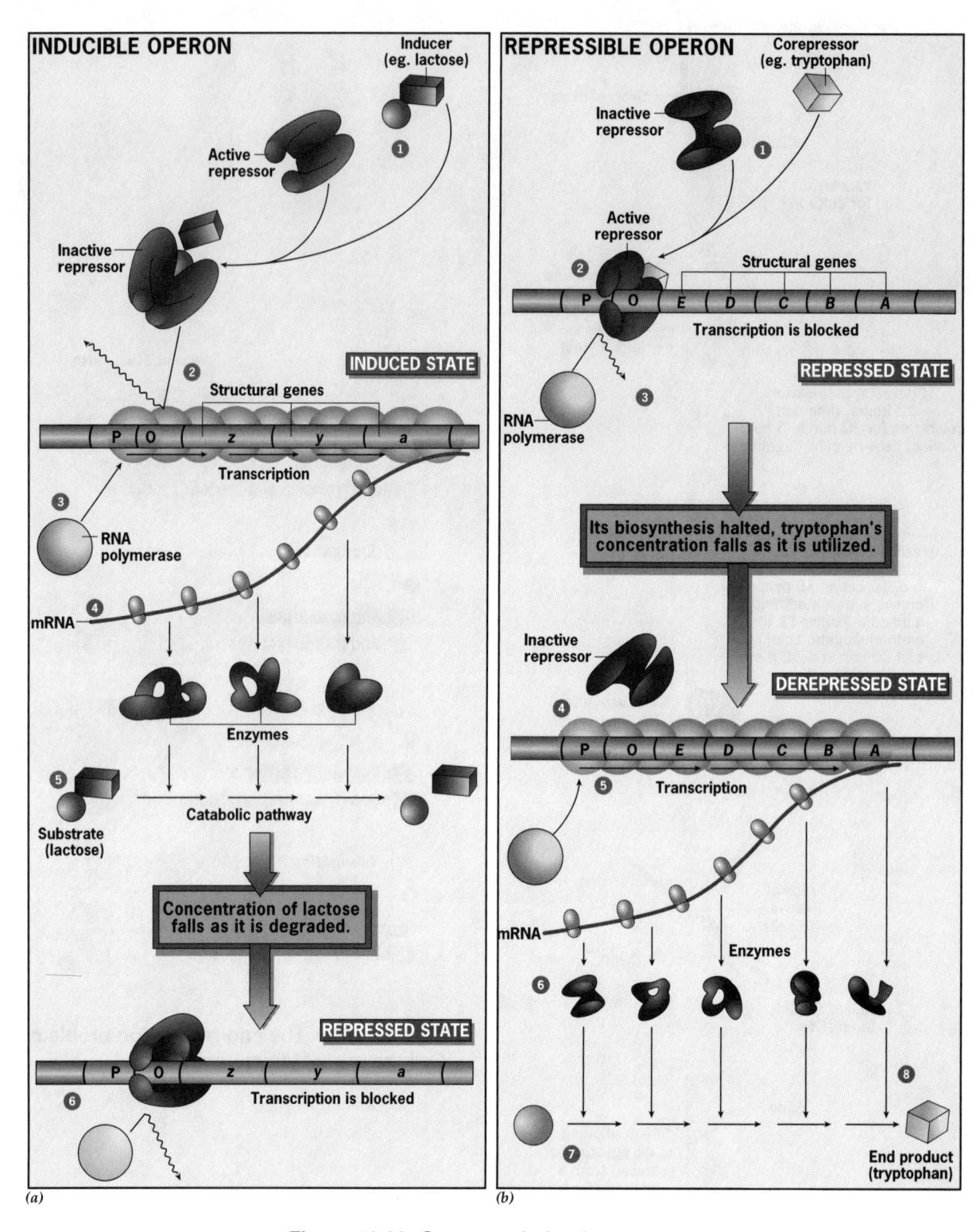

Figure 12.26 Gene regulation by operons.

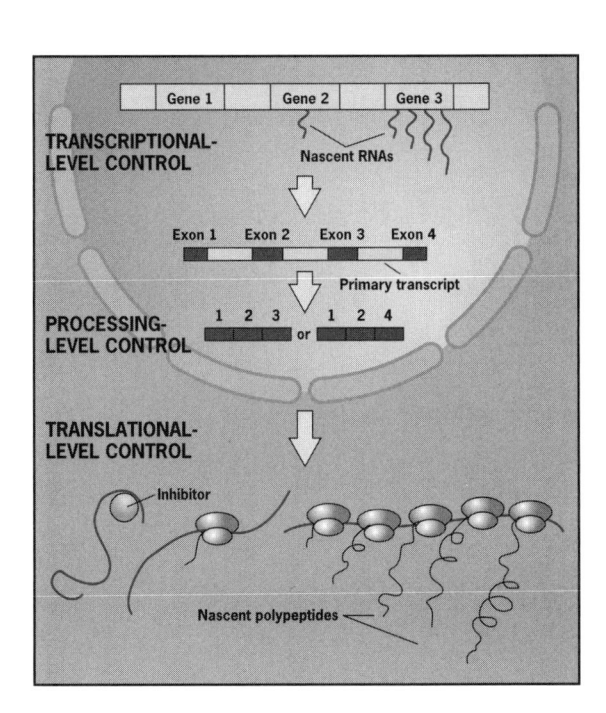

Figure 12.29 Overview of the levels of control of gene expression.

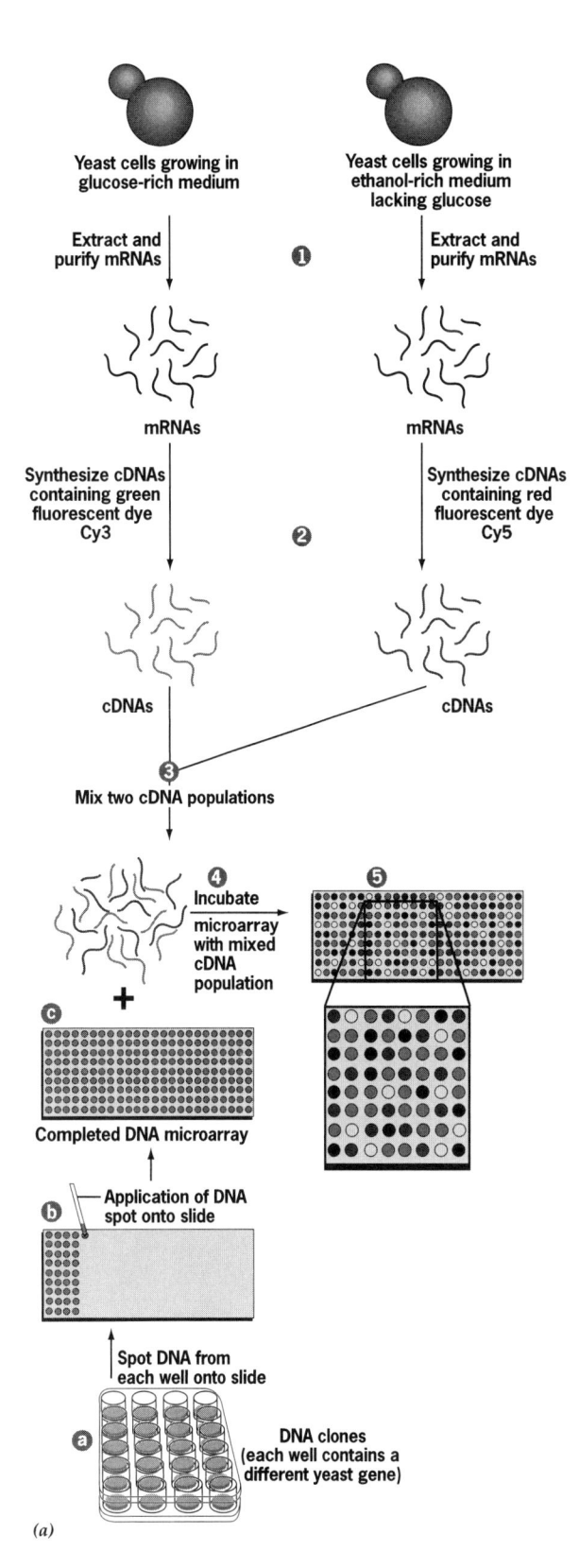

Figure 12.30a The production of DNA microarrays and their use in monitoring gene transcription.

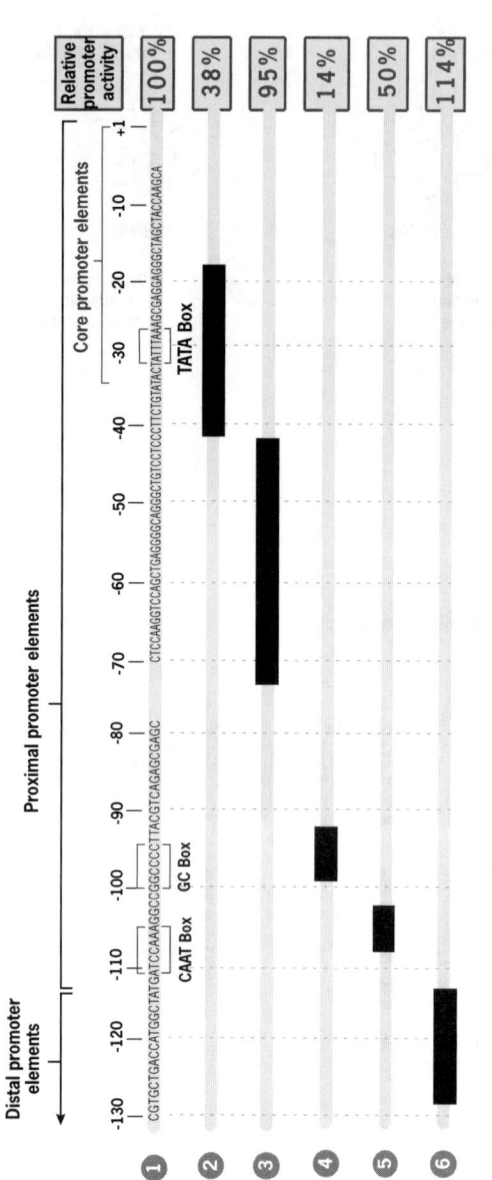

Figure 12.32 Identifying promoter sequences required for transcription.

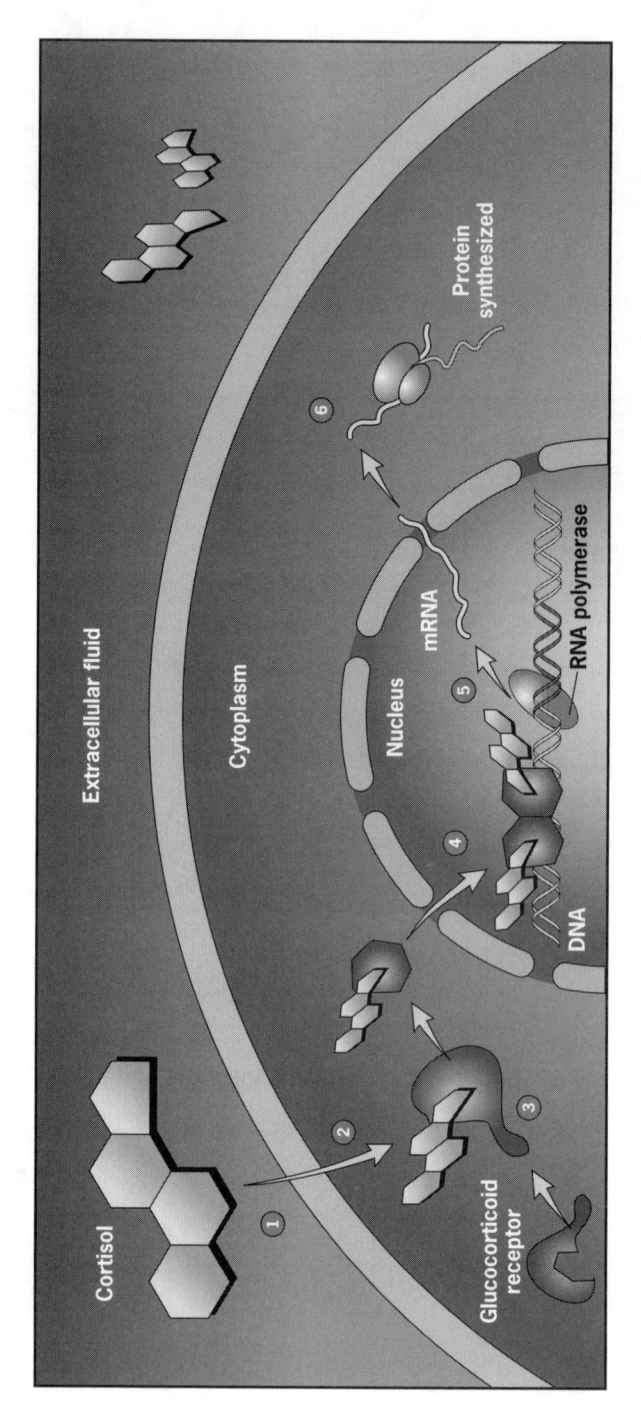

Figure 12.34 Steps in the activation of a gene by a steroid hormone, such as the glucocorticoid cortisol.

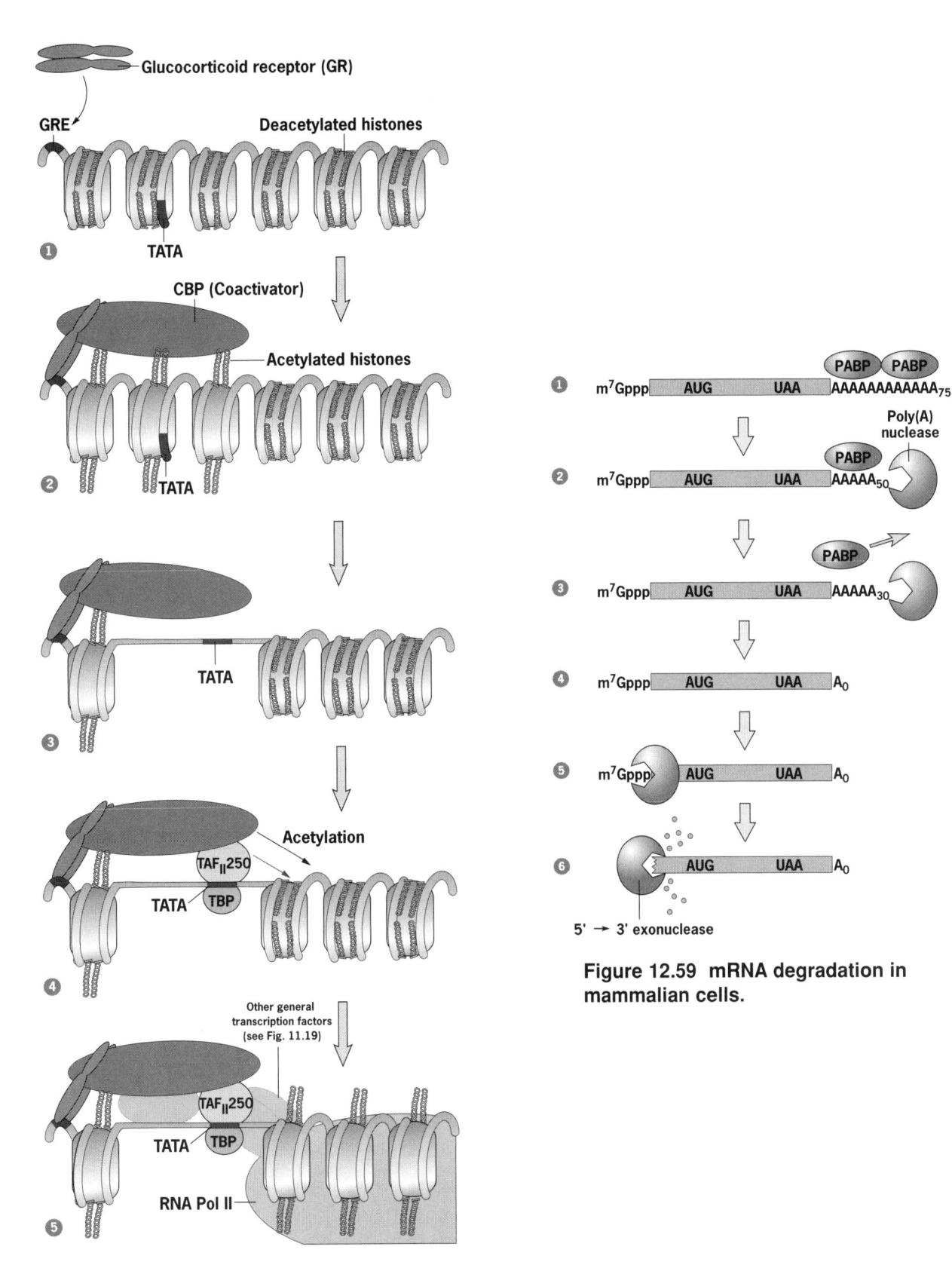

Figure 12.47 A model for the role of coactivators in regulating transcription by histone acetylation.

Figure 12.59 mRNA degradation in mammalian cells.

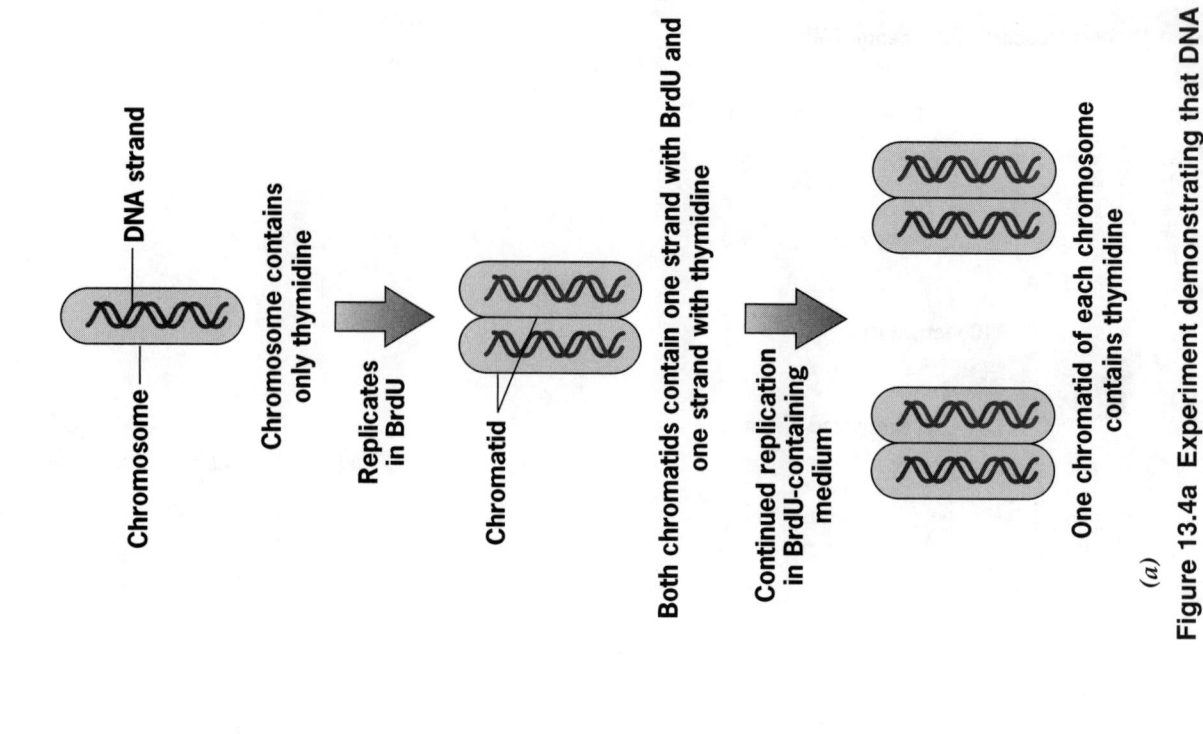

Figure 13.4a Experiment demonstrating that DNA replication in eukaryotic cells is semiconservative.

Figure 13.3a Experiment demonstrating that DNA replication in bacteria is semiconservative.

Figures 13.3a & 13.4a

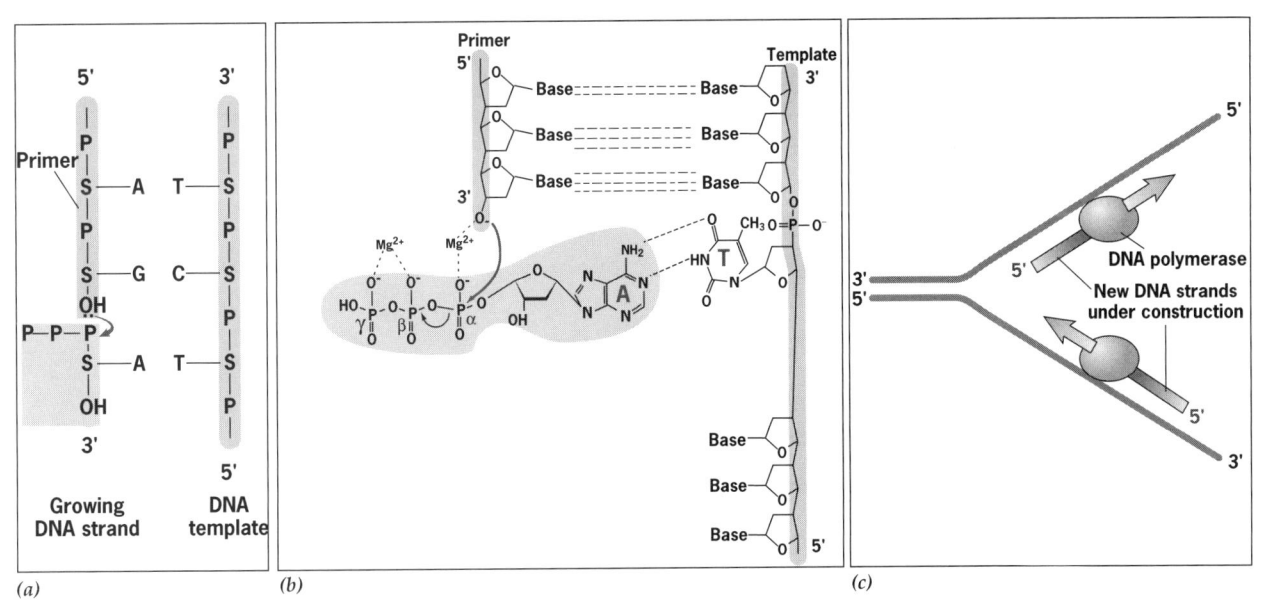

Figure 13.8 The incorporation of nucleotides onto the 3′ end of a growing strand by a DNA polymerase.

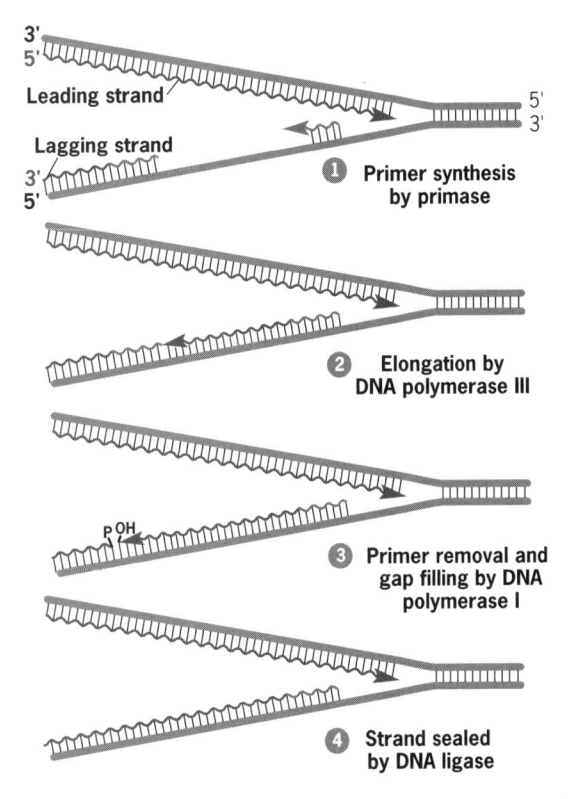

Figure 13.11 The use of short RNA fragments as removable primers in initiating synthesis of each Okazaki fragment of the lagging strand.

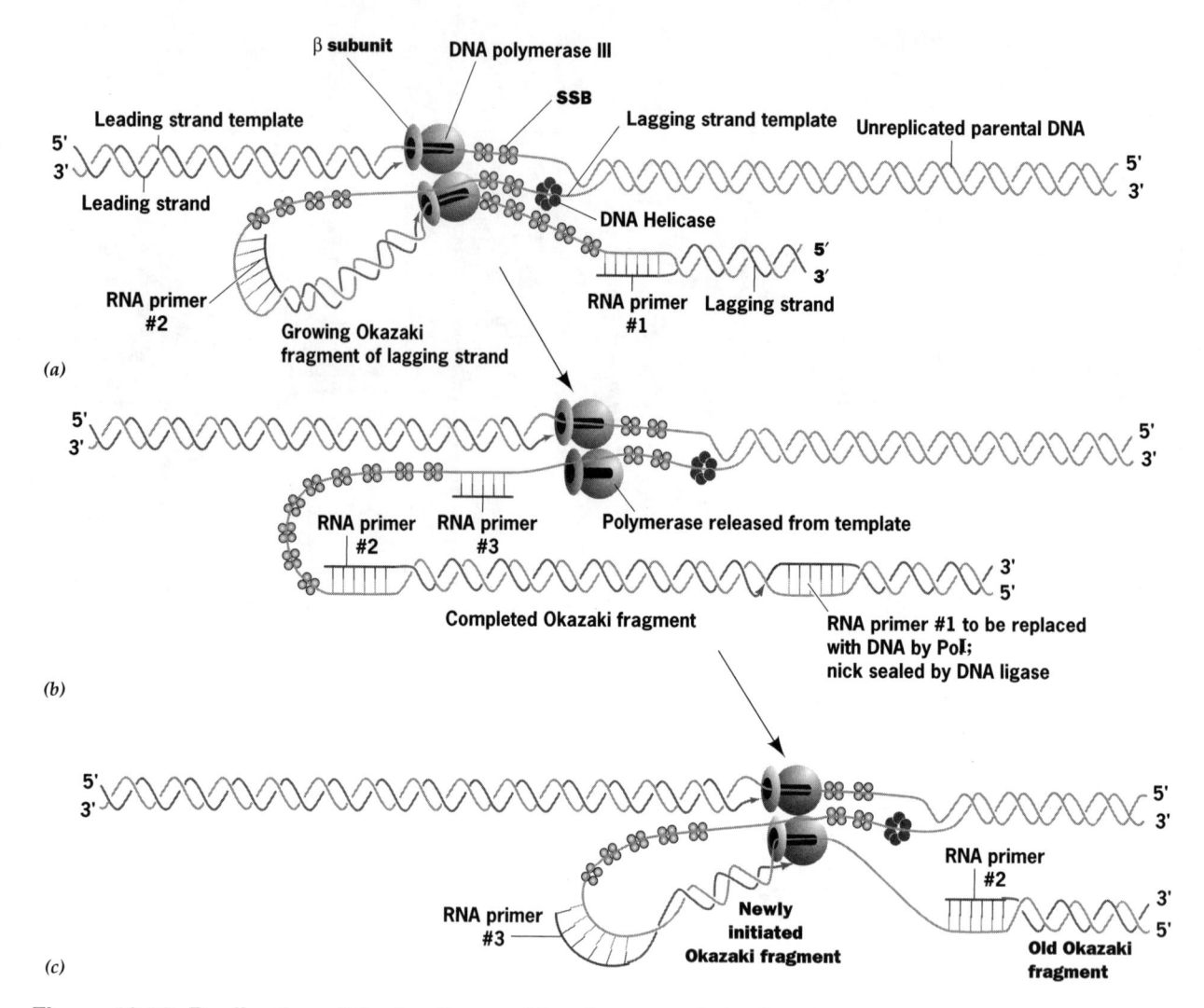

Figure 13.13 Replication of the leading and lagging strands in *E.coli* is accomplished by two DNA polymerases working together as part of a single complex.

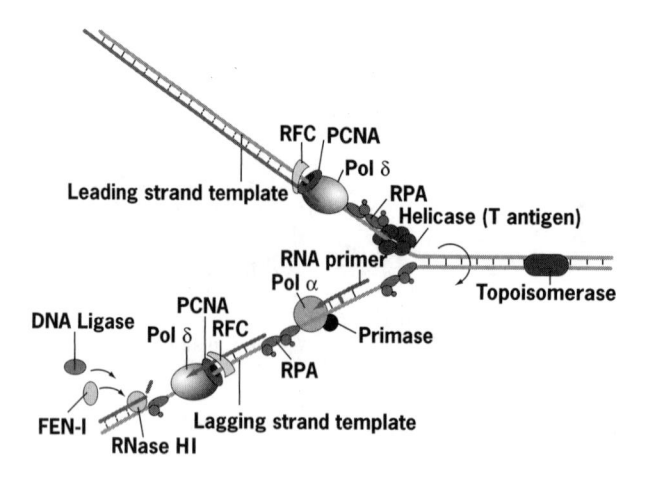

Figure 13.22 A schematic view of the major components at the eukaryotic replication fork.

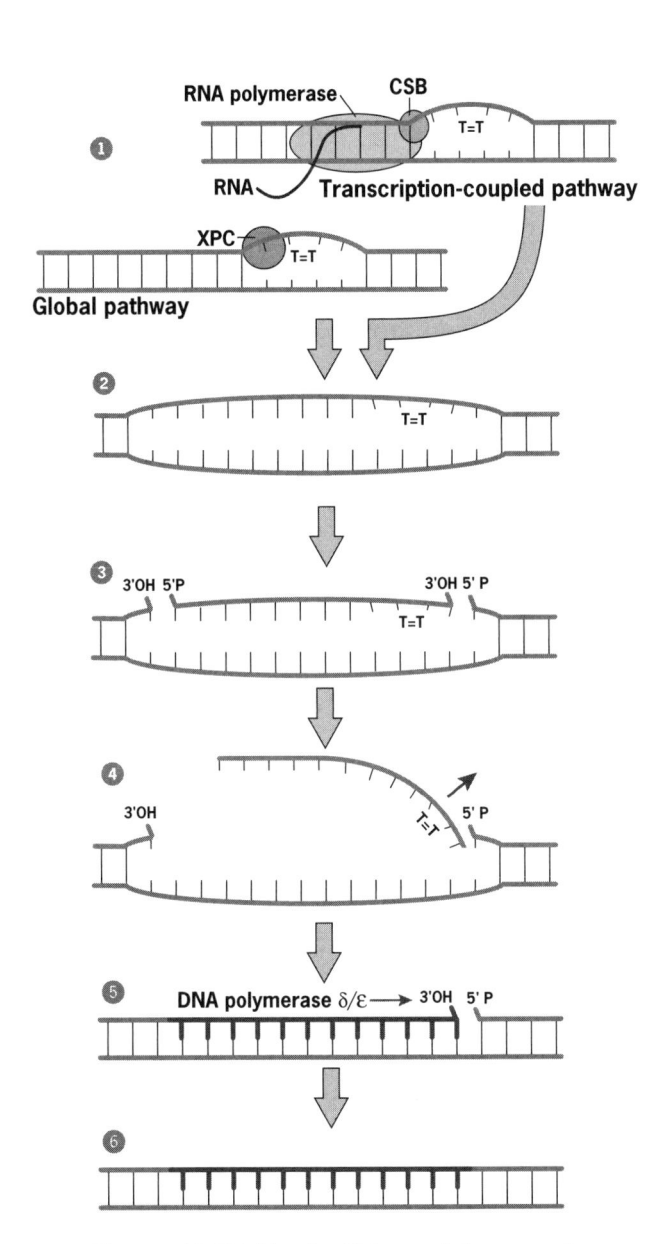

Figure 13.27 Nucleotide excision repair.

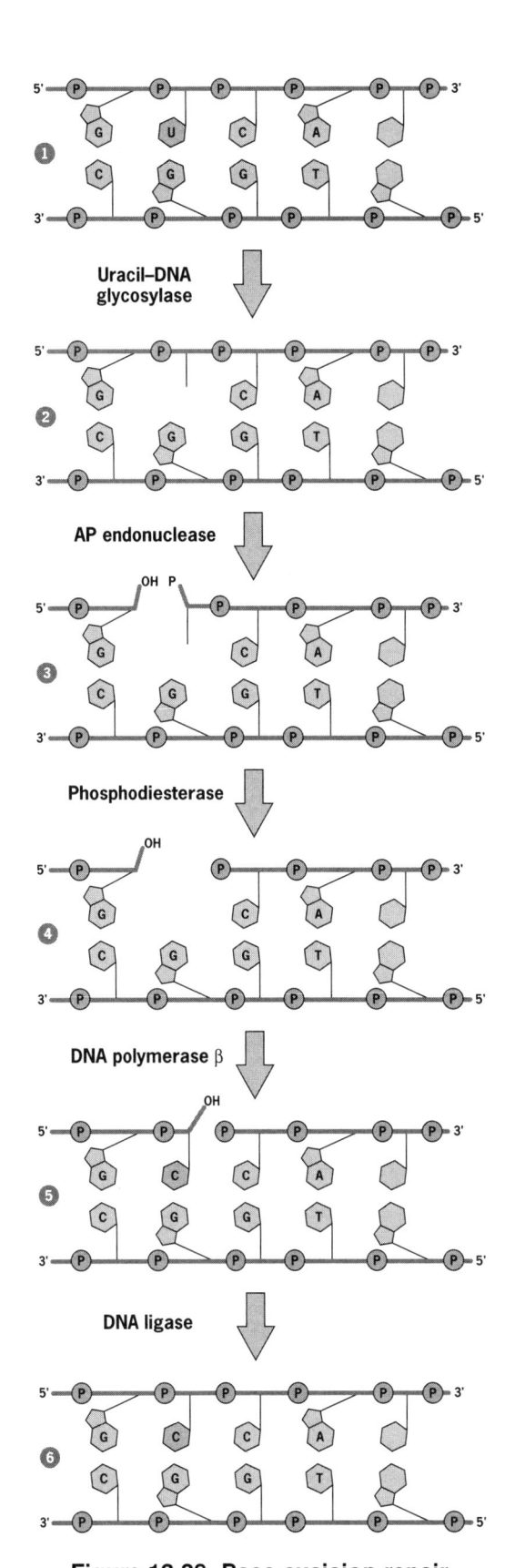

Figure 13.28 Base excision repair.

Figures 13.27 & 13.28

Figure 14.1 An overview of the eukaryotic cell cycle.

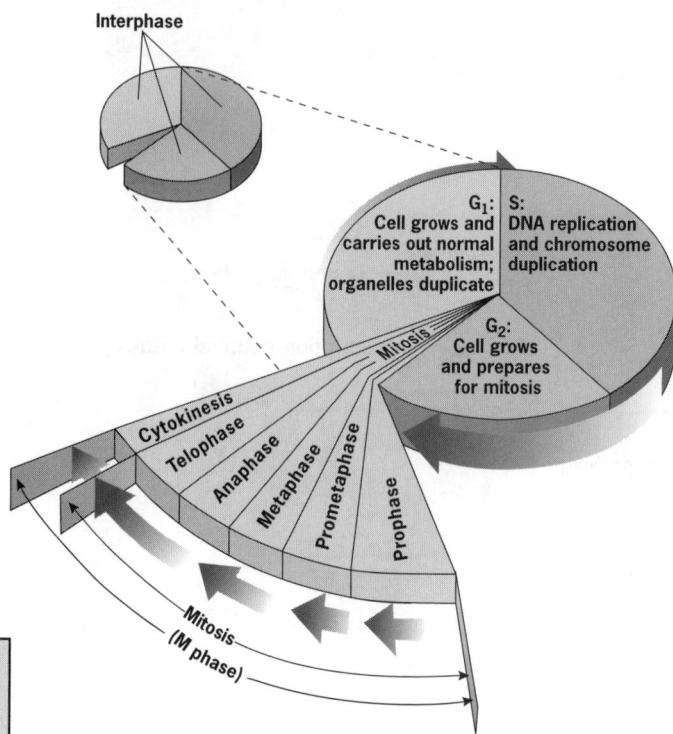

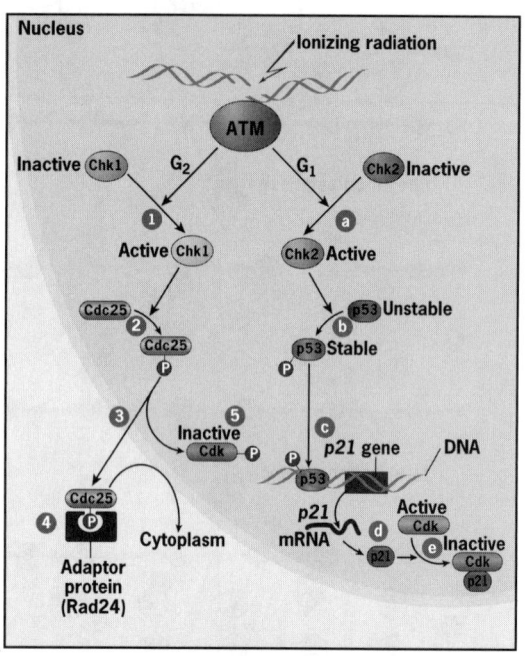

Figure 14.9 A model for the mechanism of action of a DNA-damage checkpoint.

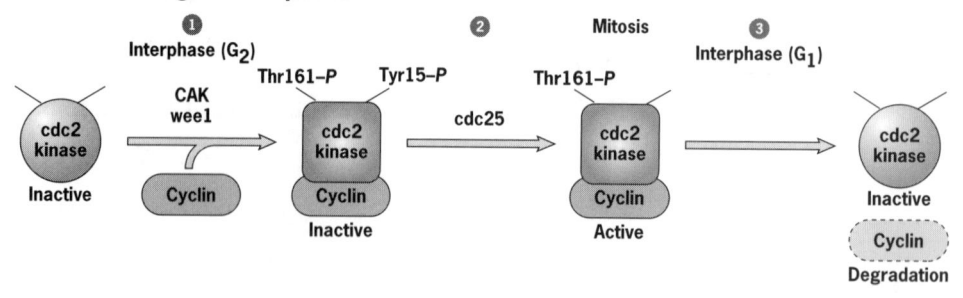

Figure 14.6 Progression through the cell cycle requires the phosphorylation and dephosphorylation of critical cdc2 residues.

Copyright © 2002 John Wiley & Sons, Inc.

Figures 14.1 & 14.6 & 14.9

Prophase

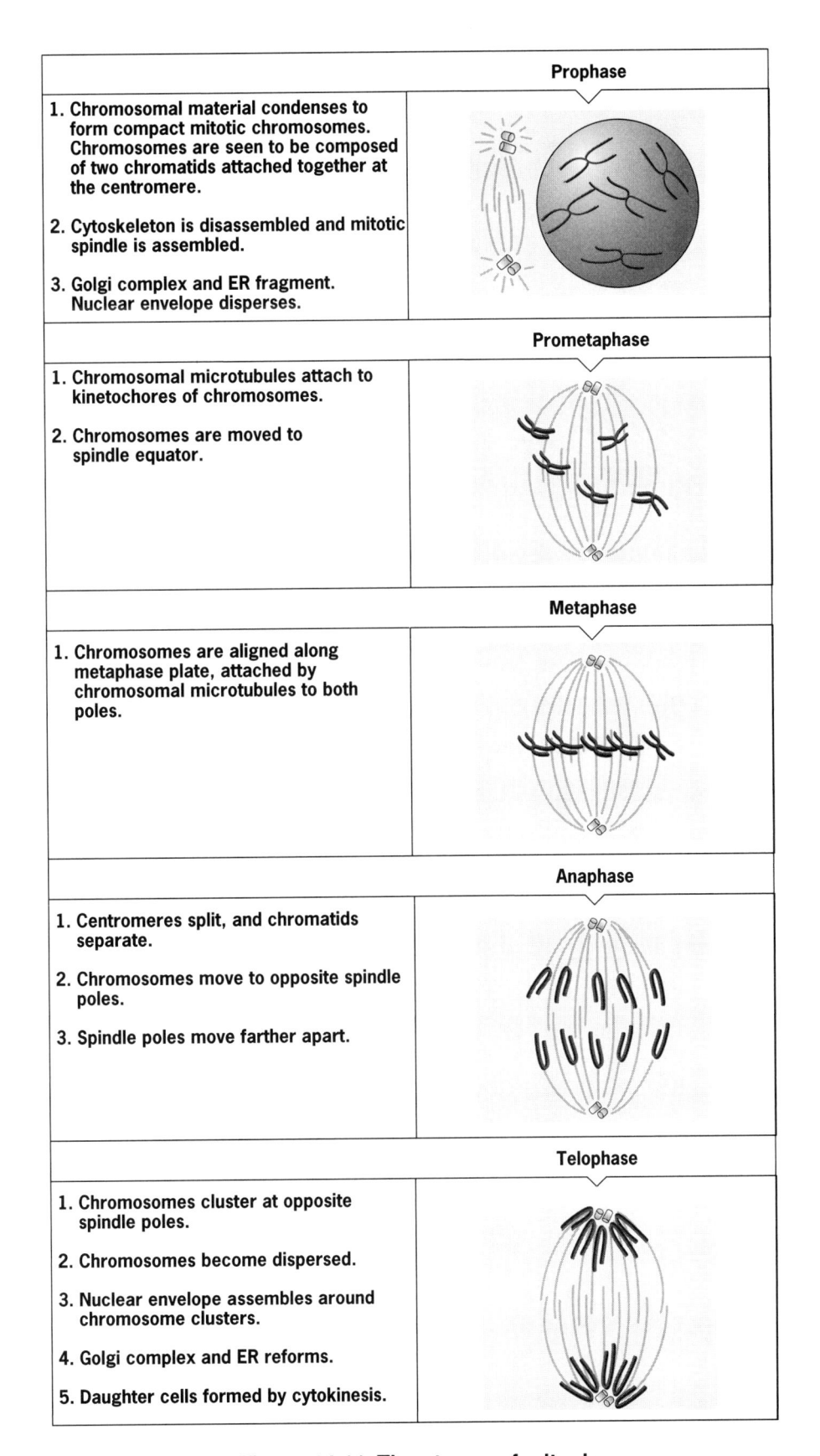

1. Chromosomal material condenses to form compact mitotic chromosomes. Chromosomes are seen to be composed of two chromatids attached together at the centromere.
2. Cytoskeleton is disassembled and mitotic spindle is assembled.
3. Golgi complex and ER fragment. Nuclear envelope disperses.

Prometaphase

1. Chromosomal microtubules attach to kinetochores of chromosomes.
2. Chromosomes are moved to spindle equator.

Metaphase

1. Chromosomes are aligned along metaphase plate, attached by chromosomal microtubules to both poles.

Anaphase

1. Centromeres split, and chromatids separate.
2. Chromosomes move to opposite spindle poles.
3. Spindle poles move farther apart.

Telophase

1. Chromosomes cluster at opposite spindle poles.
2. Chromosomes become dispersed.
3. Nuclear envelope assembles around chromosome clusters.
4. Golgi complex and ER reforms.
5. Daughter cells formed by cytokinesis.

Figure 14.11 The stages of mitosis.

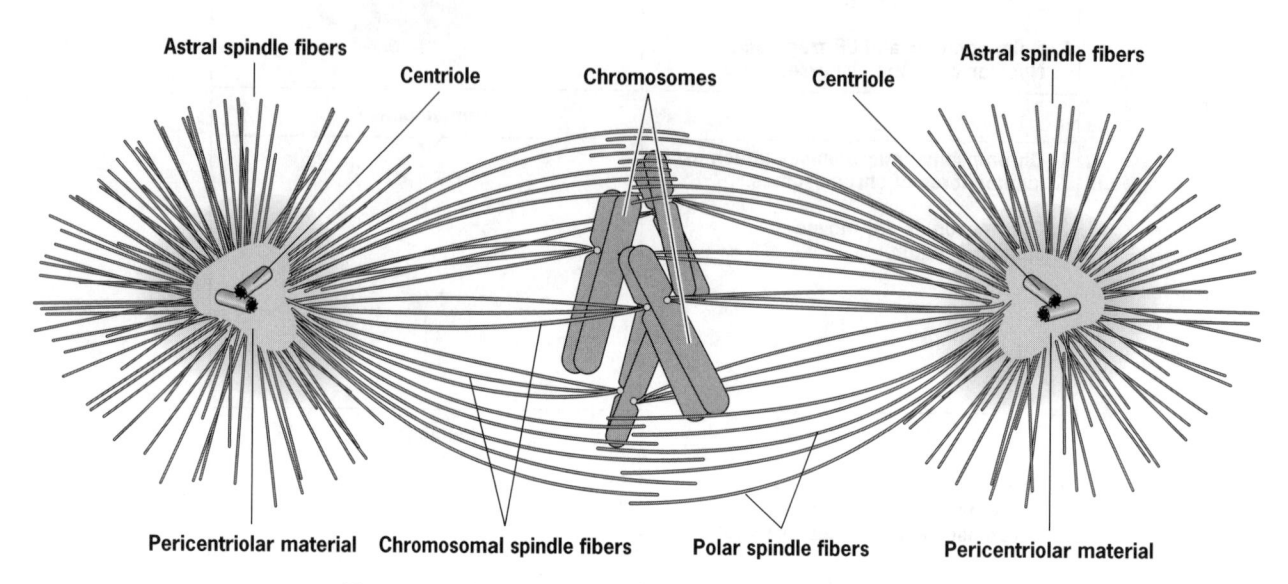

Figure 14.23 The mitotic spindle of an animal cell.

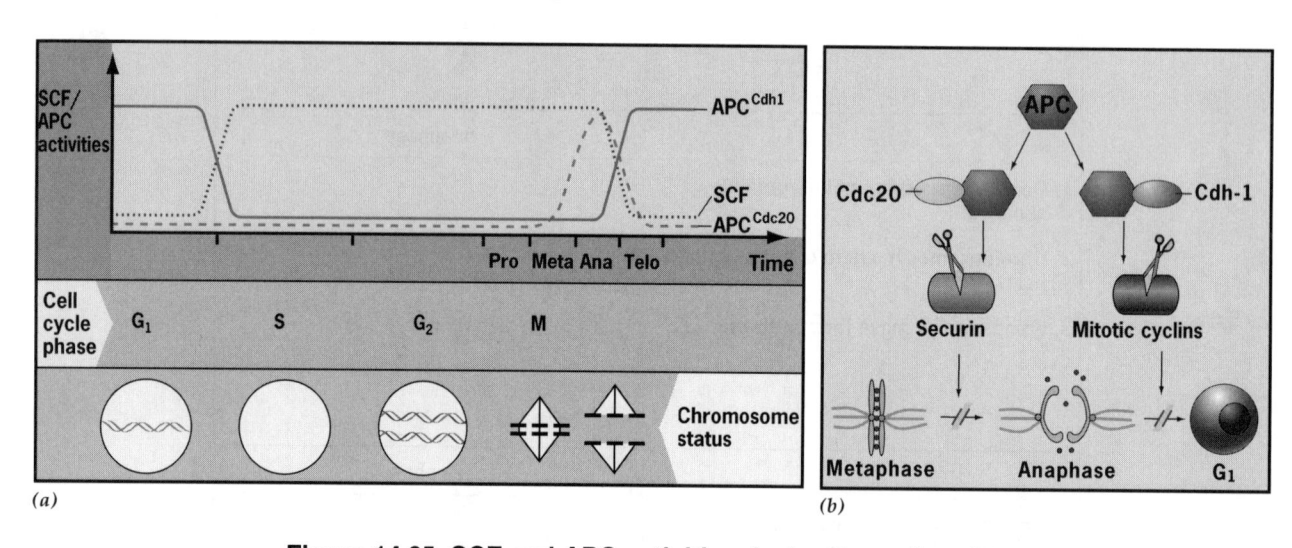

Figure 14.25 SCF and APC activities during the cell cycle.

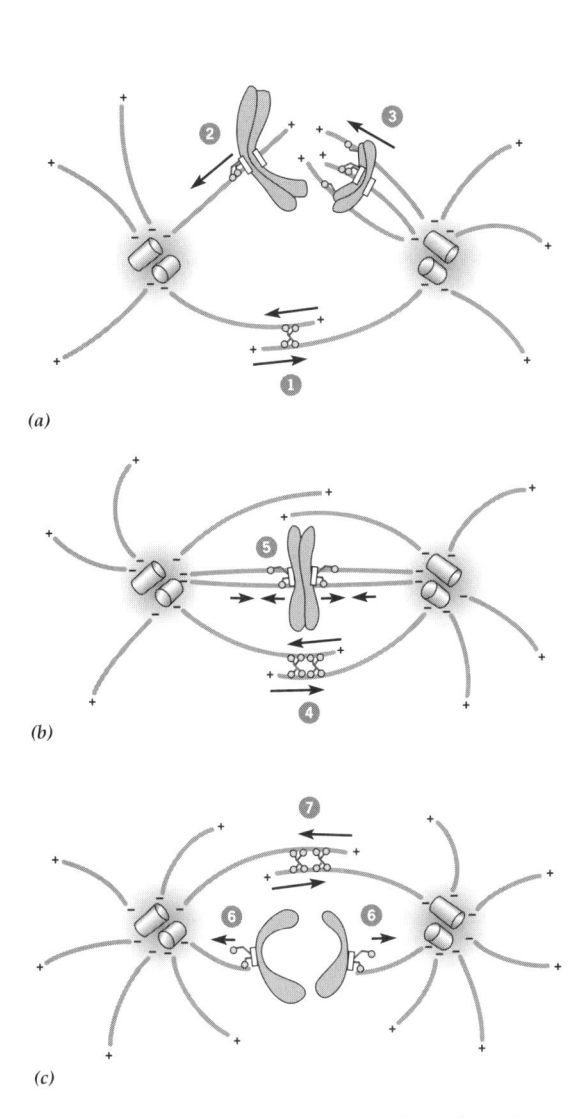

Figure 14.30 Proposed activity of motor proteins during mitosis.

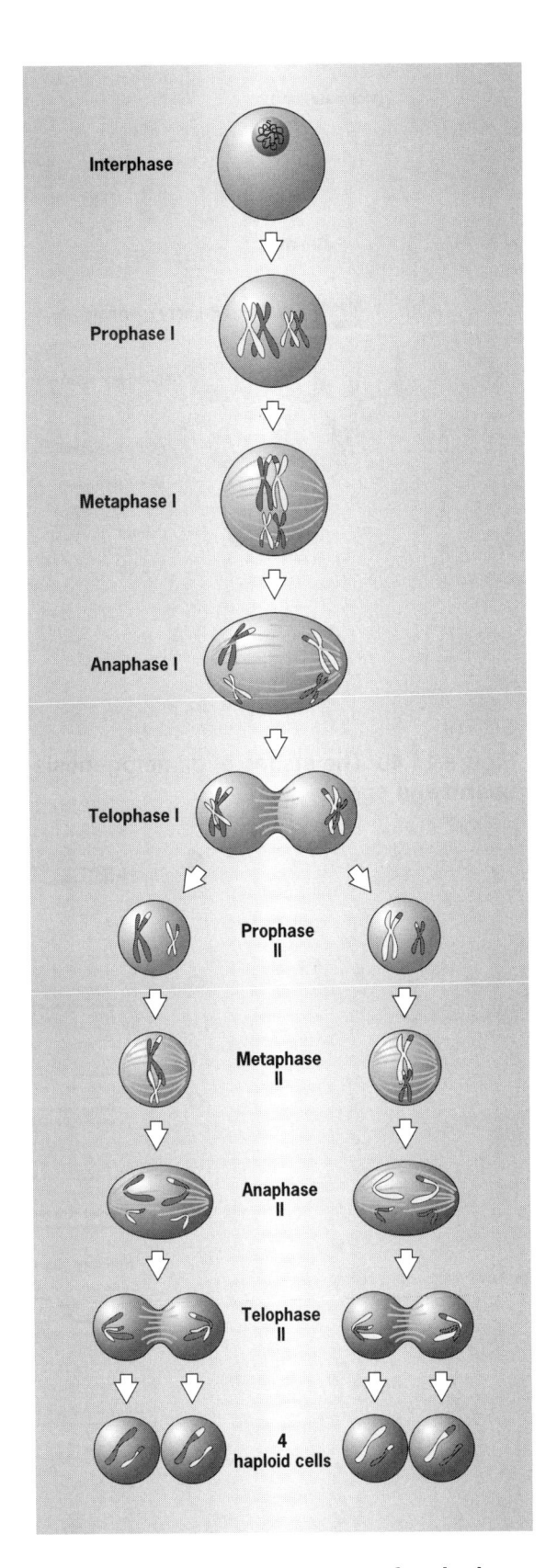

Figure 14.38 The stages of meiosis.

Figures 14.30 & 14.38

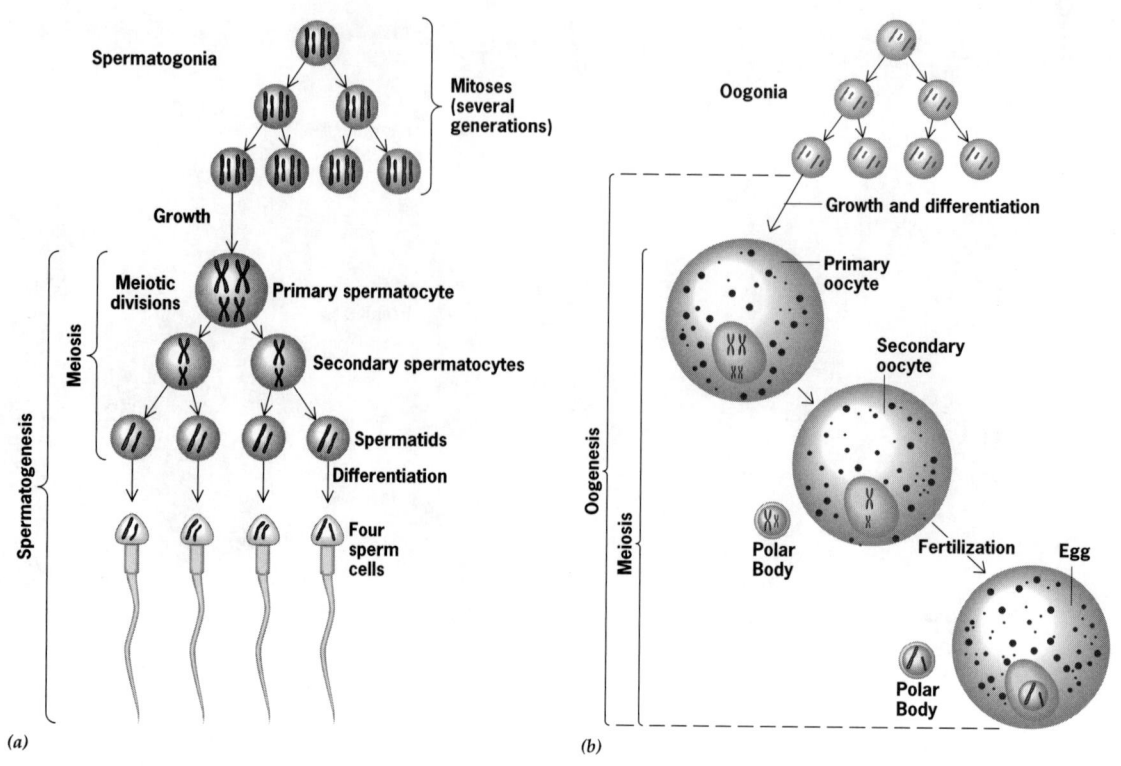

Figure 14.40 The stages of gametogenesis in vertebrates: a comparison between the formation of sperm and eggs.

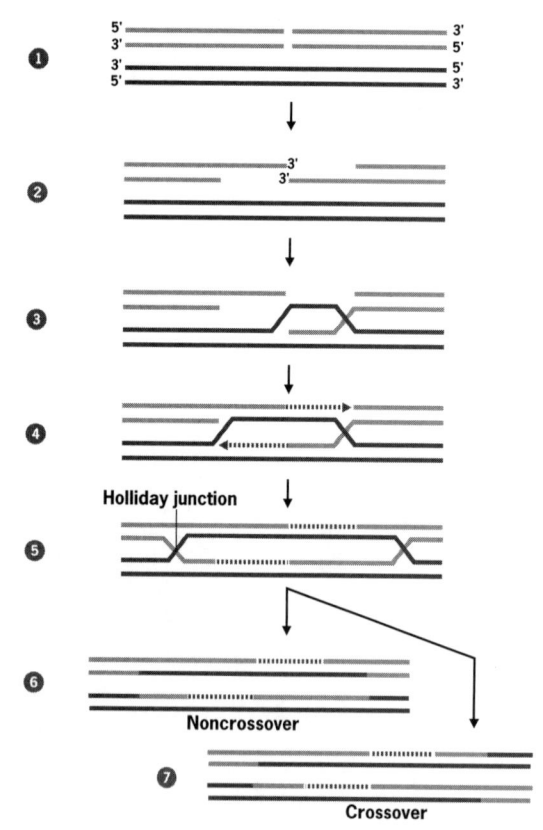

Figure 14.47 A proposed mechanism for genetic recombination initiated by double-strand breaks.

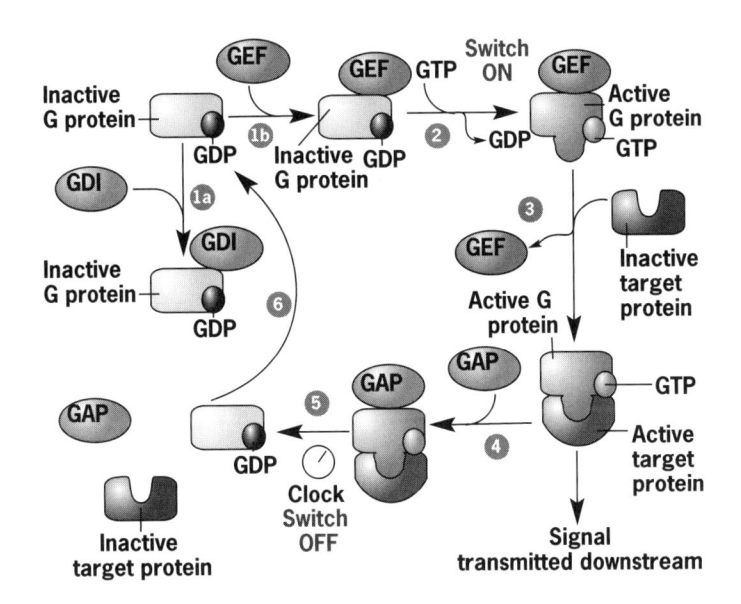

Figure 15.3b The structure of a G protein and the G protein cycle.

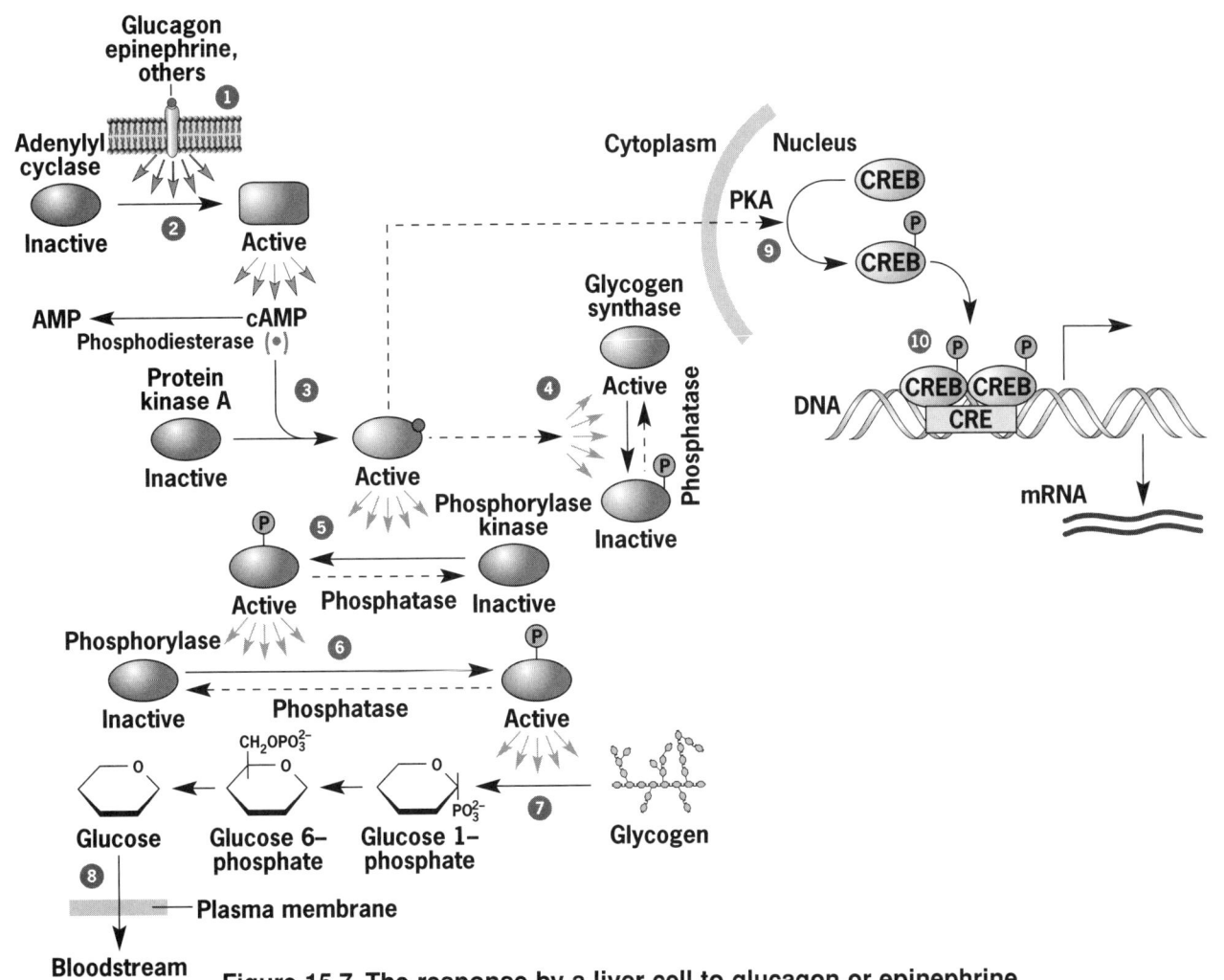

Figure 15.7 The response by a liver cell to glucagon or epinephrine.

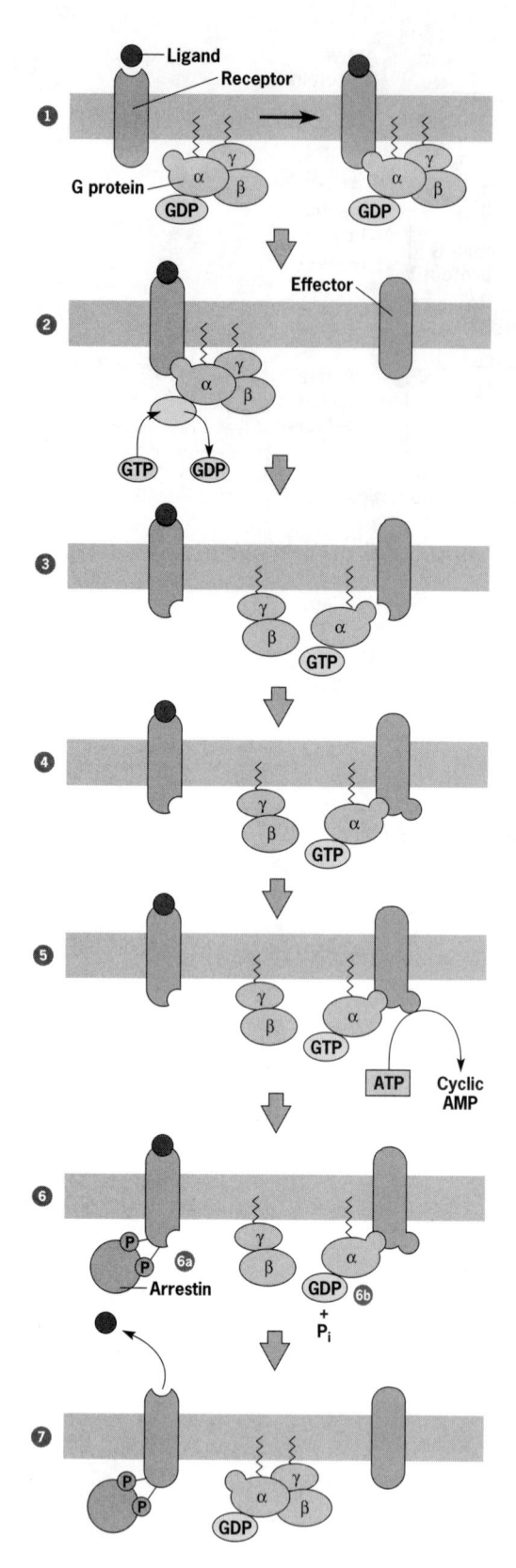

Figure 15.10 The mechanism of receptor-mediated activation (or inhibition) of effectors by means of heterotrimeric G proteins.

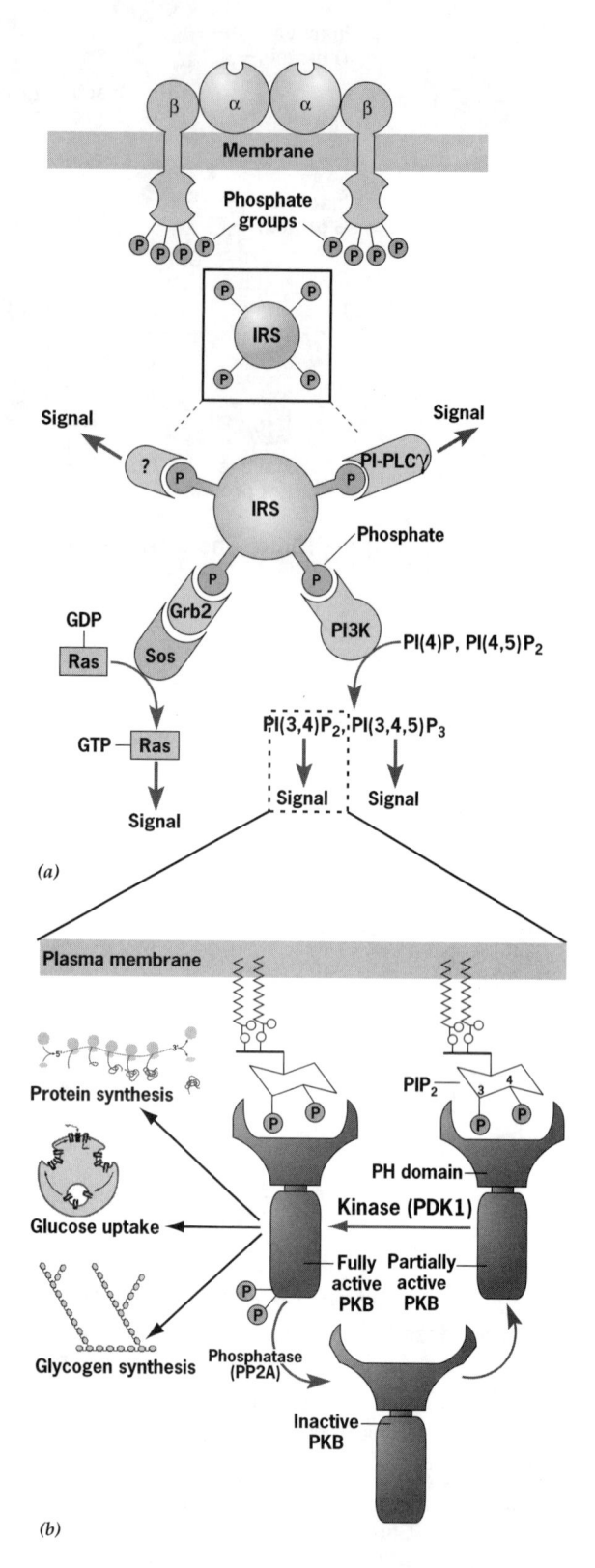

Figure 15.21 The role of tryosine-phosphorylated IRS in activating a variety of signaling pathways.

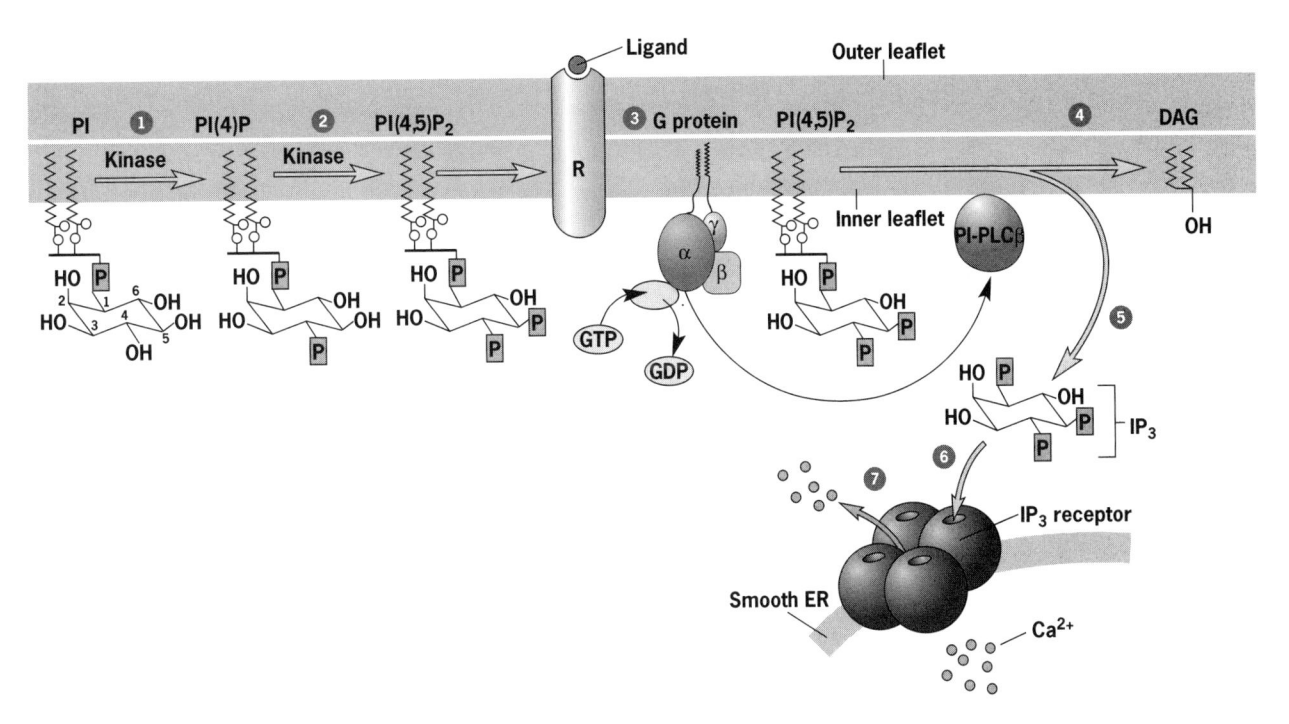

Figure 15.12 The generation of second messengers as a result of ligand-induced breakdown of phosphatidylinositol (PI) in the lipid bilayer.

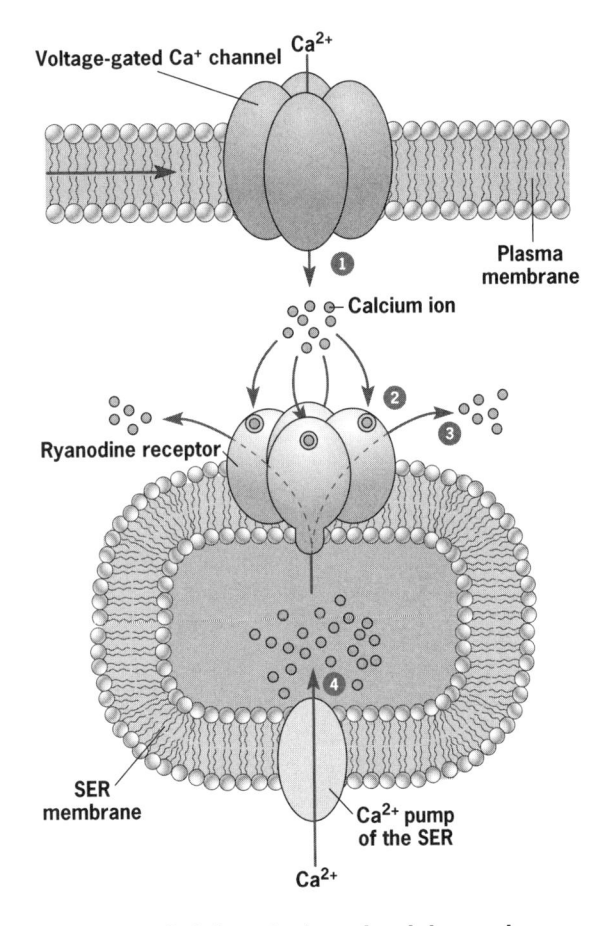

Figure 15.15 Calcium-induced calcium release.

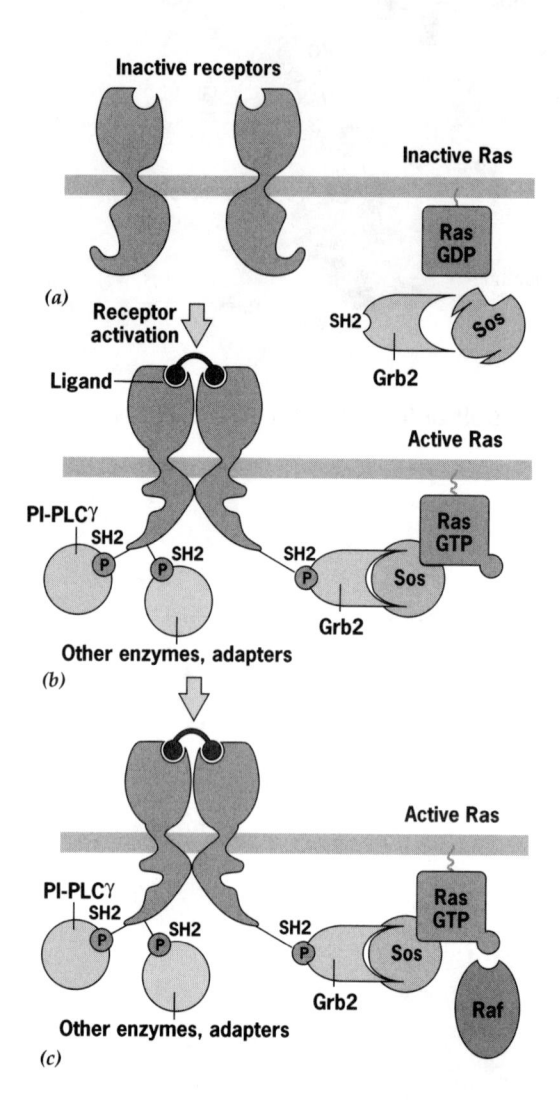

Figure 15.23 Steps in the activation of Ras by RTKs.

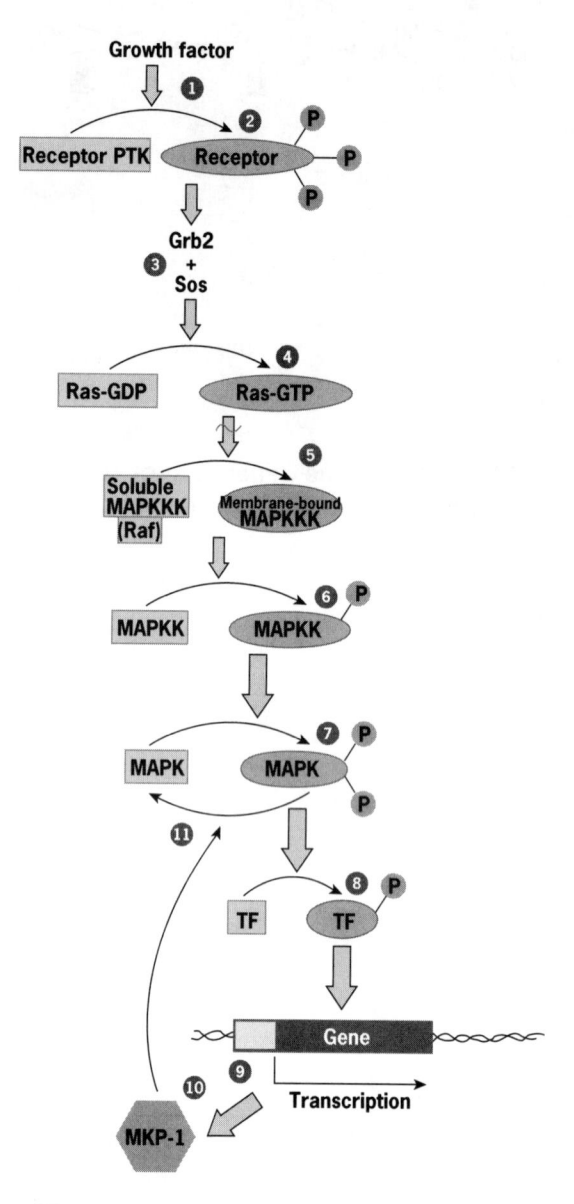

Figure 15.25 The steps of a generalized MAP kinase cascade.

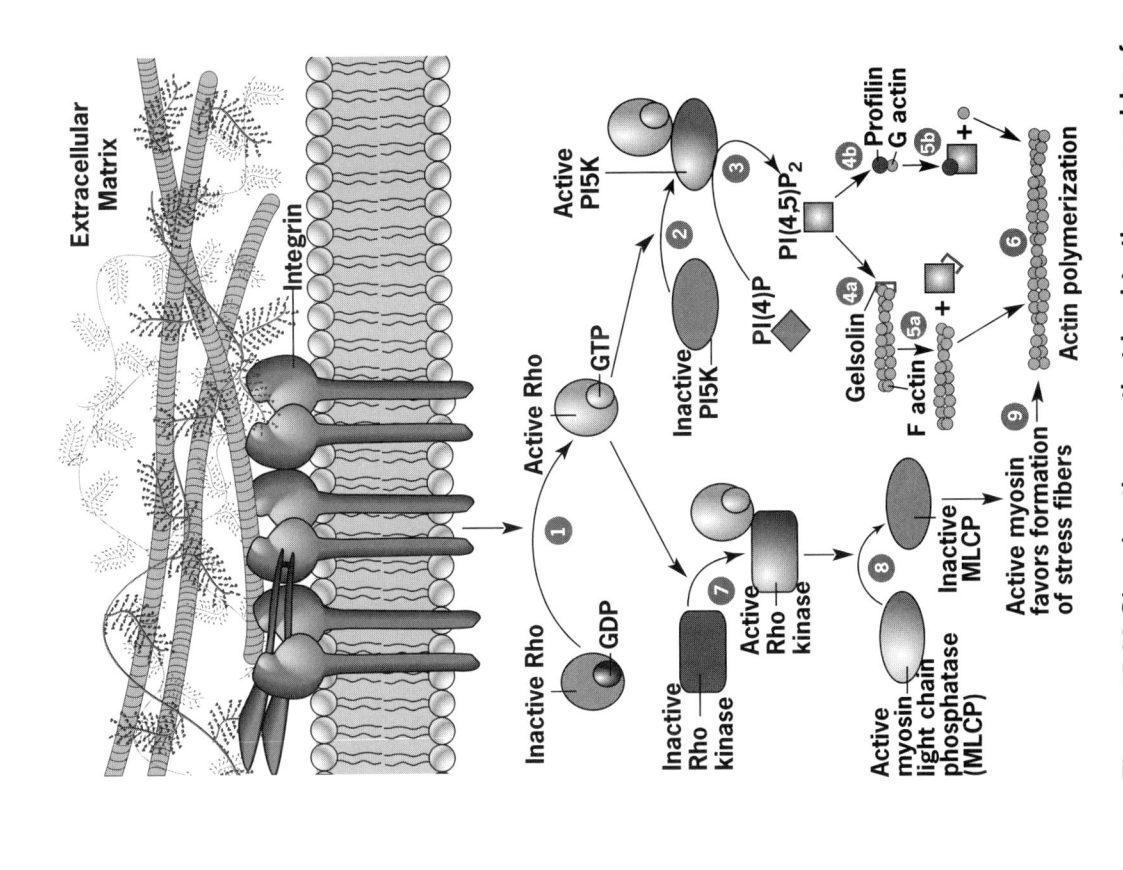

Figure 15.30 Signal pathways that lead to the assembly of the stress fibers of a focal adhesion.

Figure 15.28 Schematic model of the protein-protein interactions of a focal adhesion complex.

Copyright © 2002 John Wiley & Sons, Inc.

Figures 15.28 & 15.30

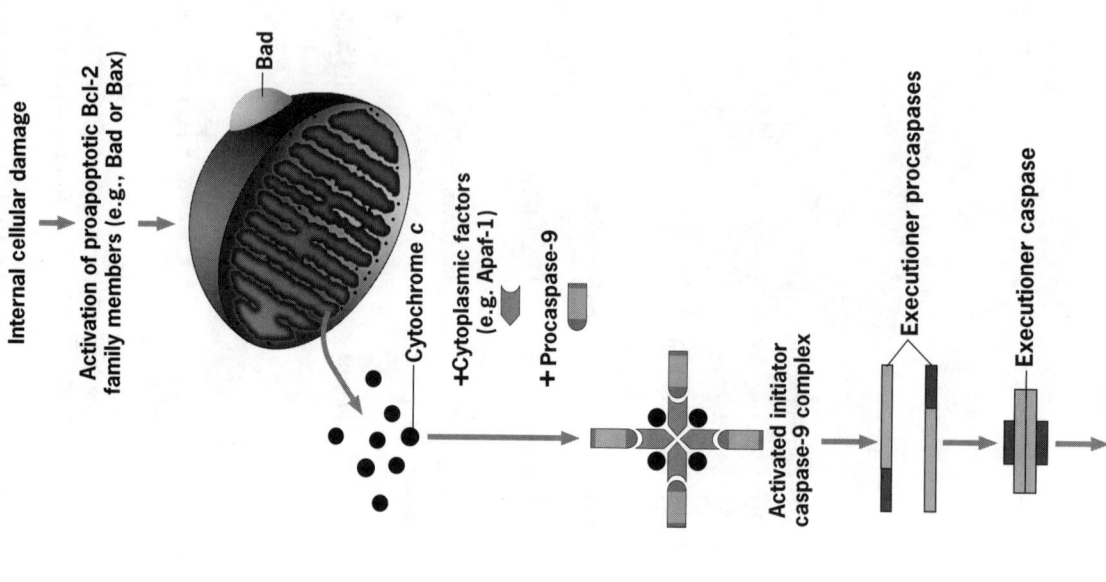

Figure 15.37 The mitochondria-mediated pathway of apoptosis.

Figure 15.36 The receptor-mediated pathway of apoptosis.

Figures 15.36 & 15.37

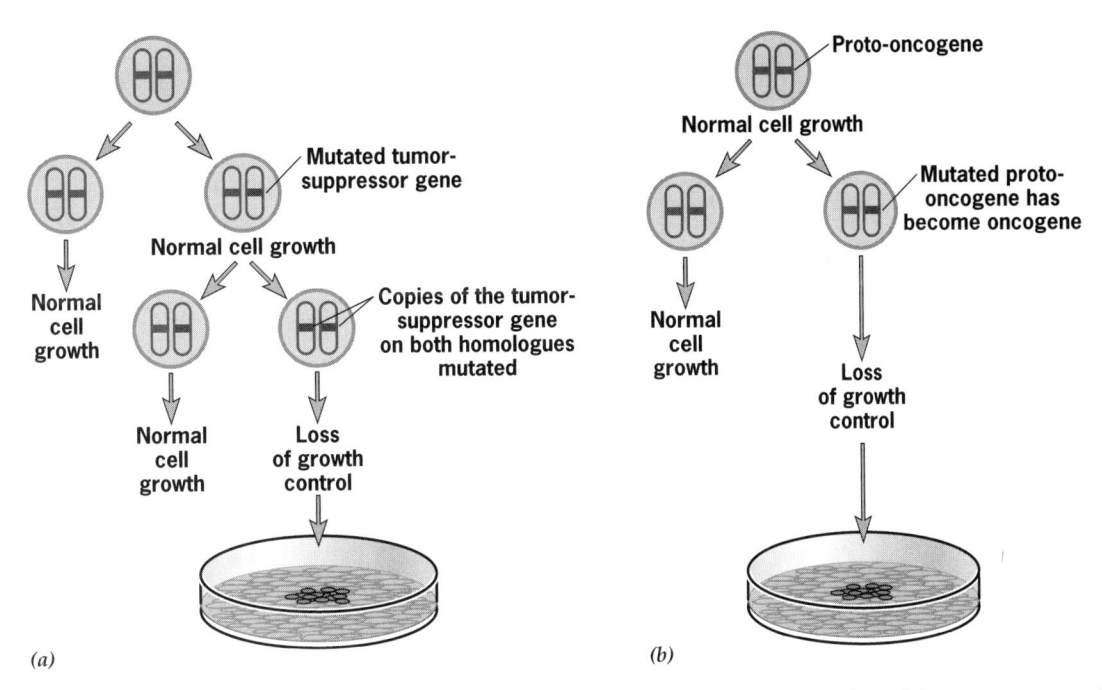

Figure 16.11 Contrasting effects of mutation in tumor-suppressor genes (a) and in oncogenes (b).

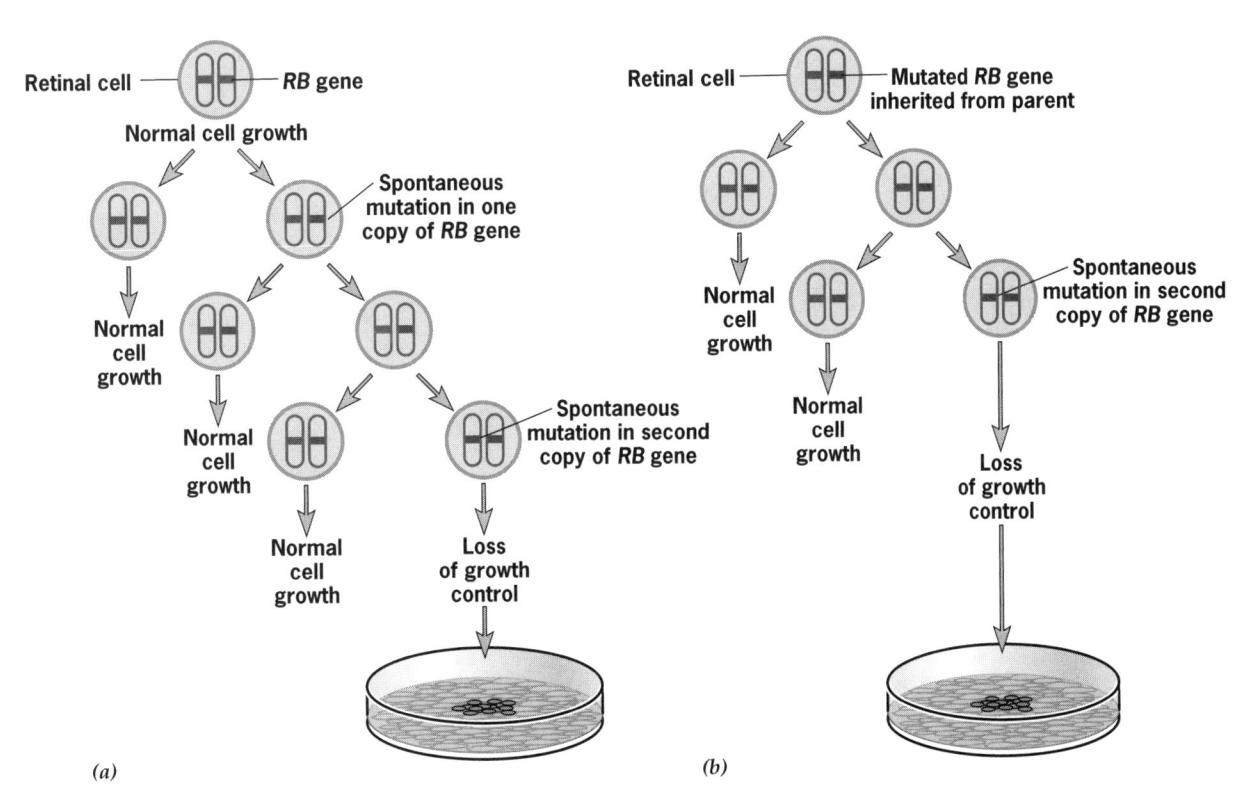

Figure 16.13 Mutations in the *RB* gene that can lead to retinoblastoma.

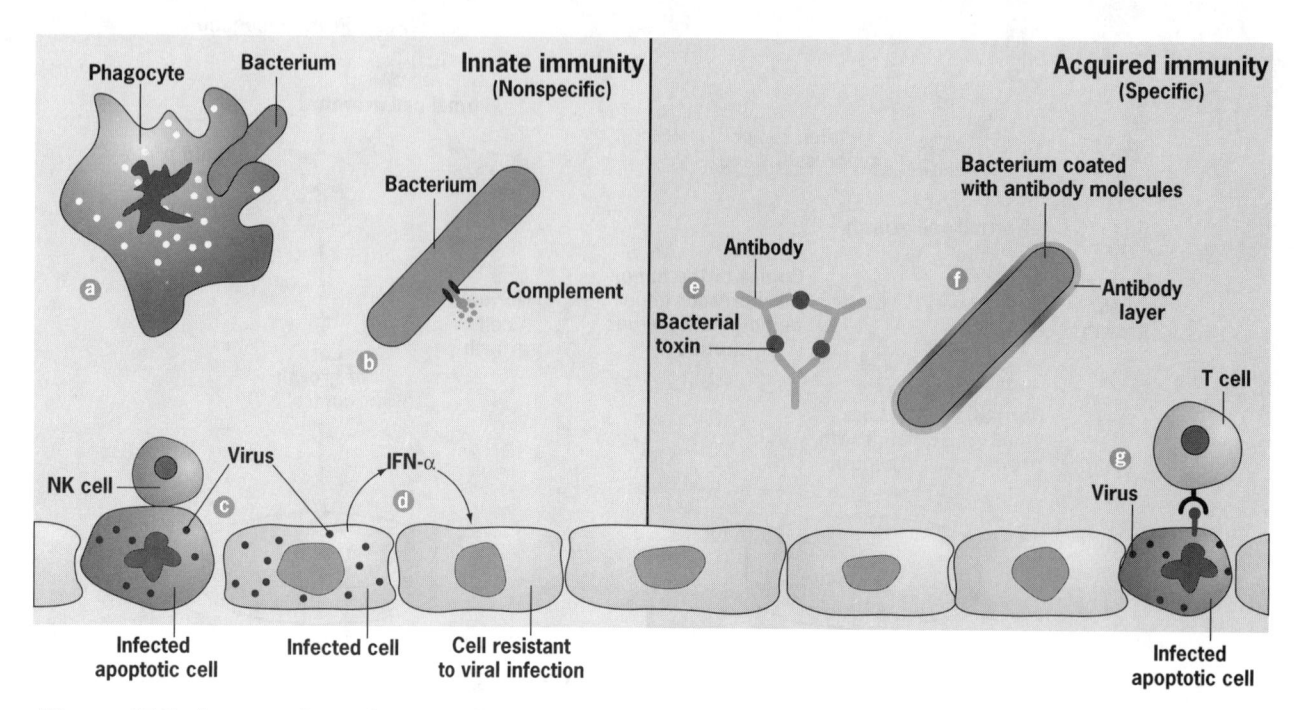

Figure 17.2 An overview of some of the mechanisms by which the immune system rids the body of invading pathogens.

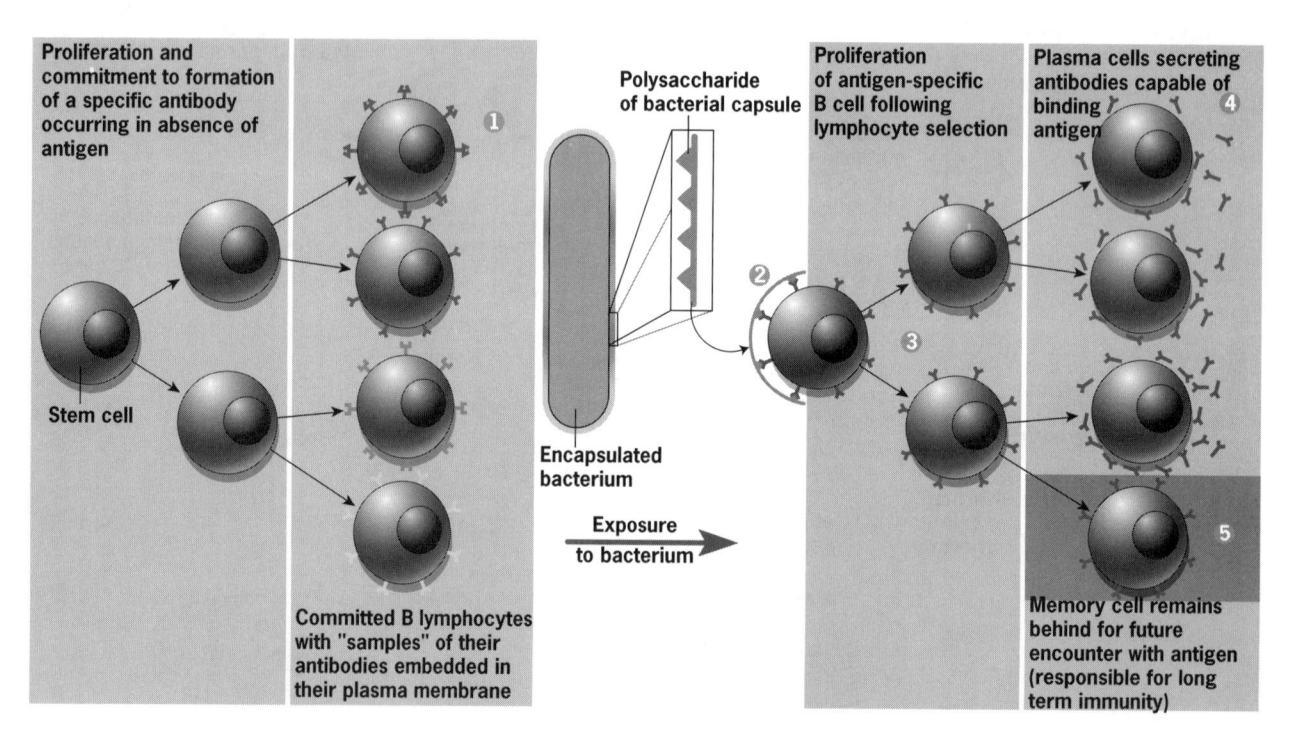

Figure 17.5 The clonal selection of B cells by a thymus-independent antigen.

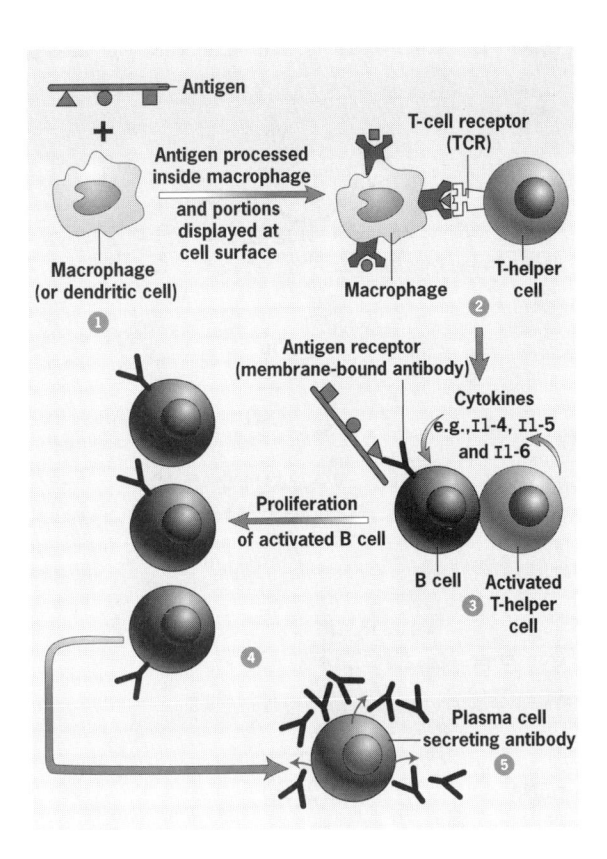

Figure 17.9 Highly simplified, schematic drawing showing the role of T_H cells in antibody formation.

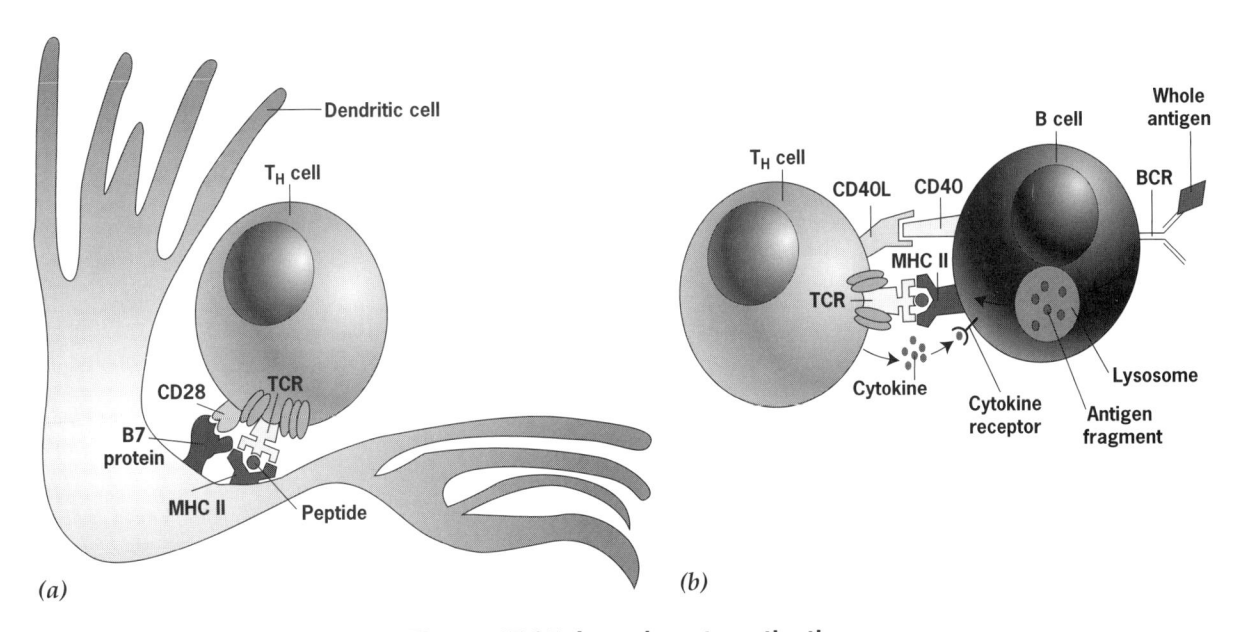

Figure 17.25 Lymphocyte activation.

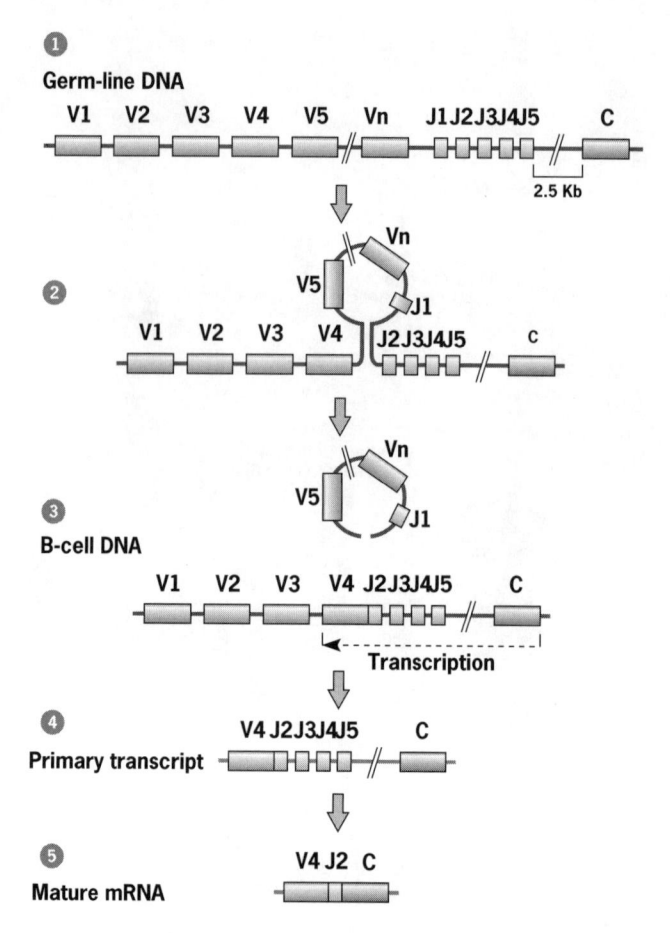

Figure 17.15 DNA rearrangements that lead to the formation of a functional gene that encodes an immunoglobulin κ chain.

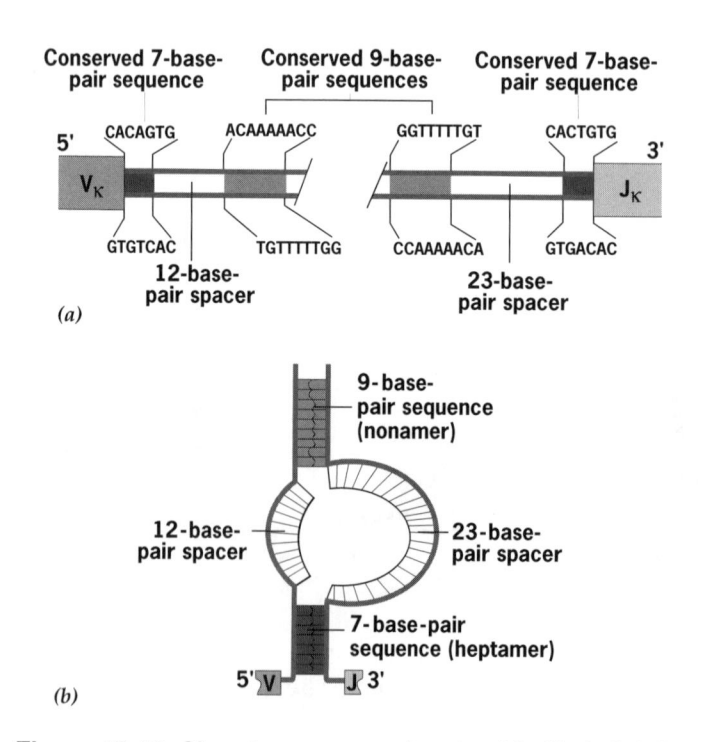

Figure 17.16 Signal sequences involved in V_κ-j_κ joining.

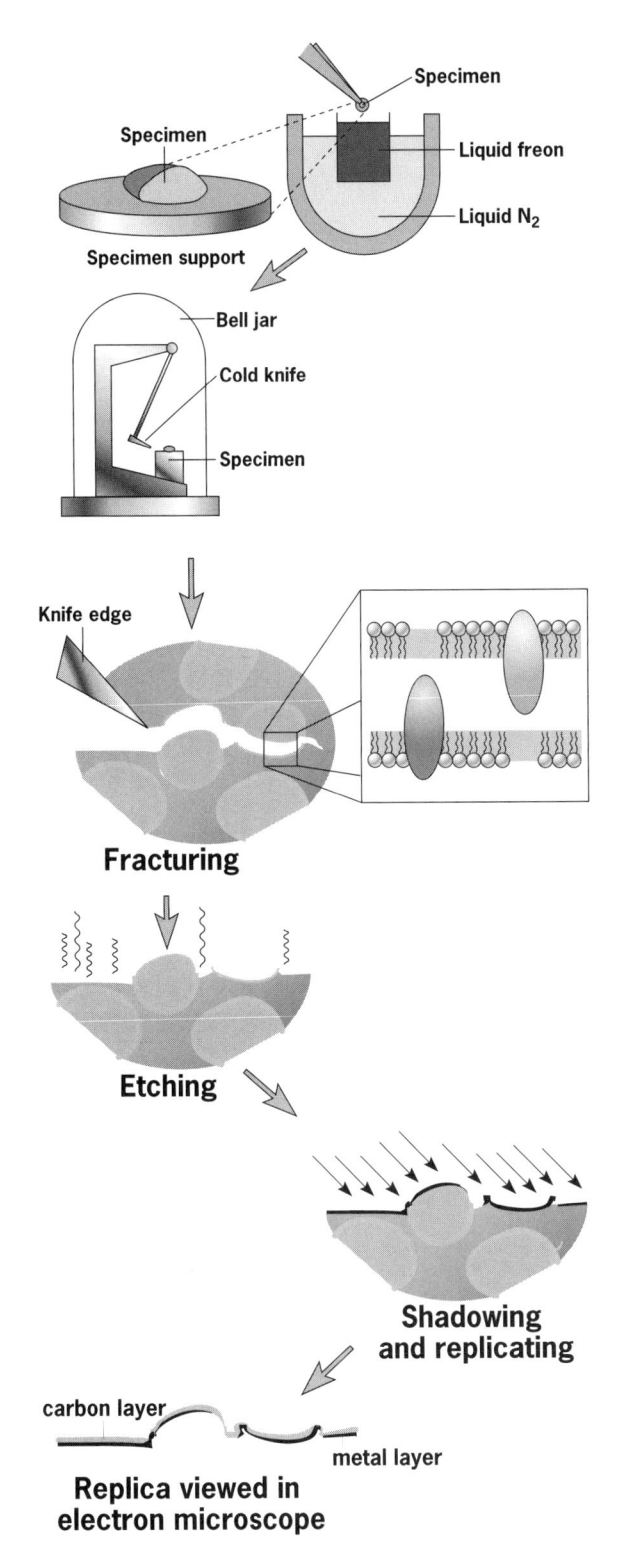

Figure 18.16 Procedure for the formation of freeze-fracture replicas as described in the text.

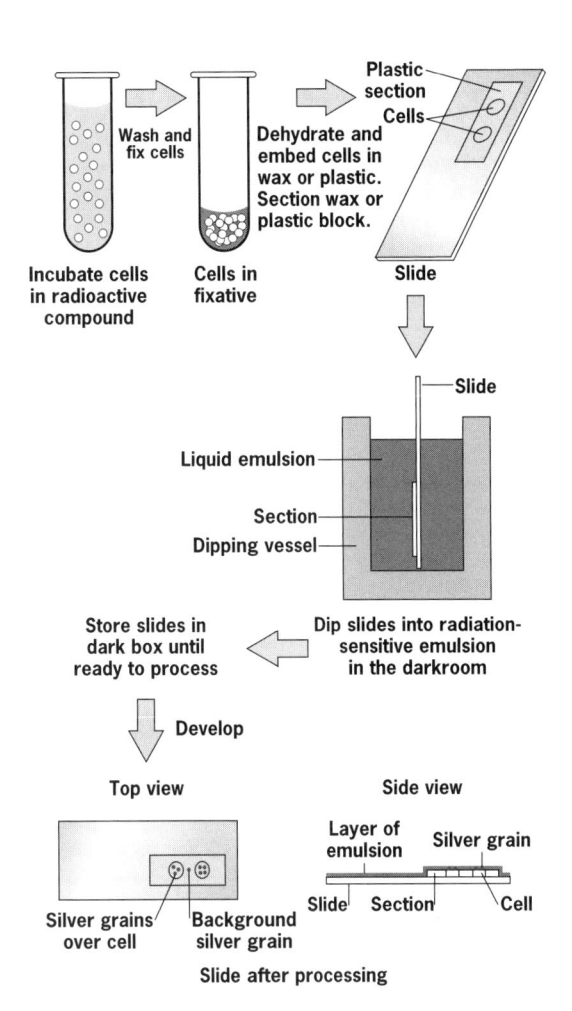

Figure 18.20 Step-by-step procedure for the preparation of an autoradiograph.

Figures 18.16 & 18.20

Figure 18.23 Cell fractionation.

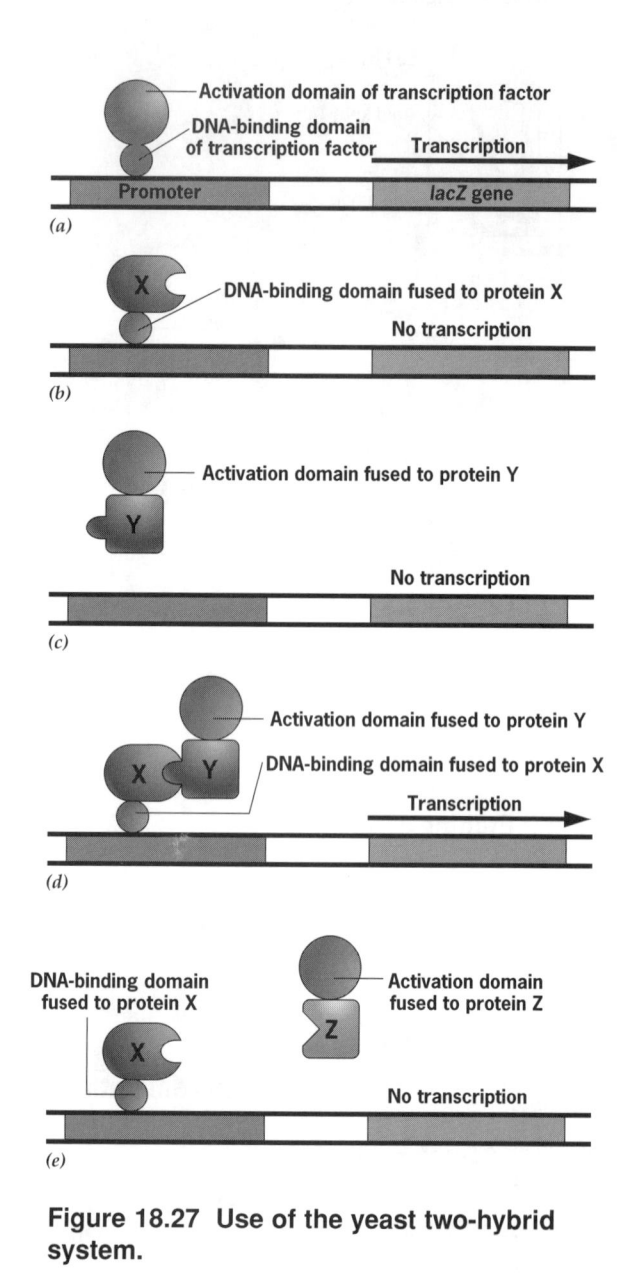

Figure 18.27 Use of the yeast two-hybrid system.

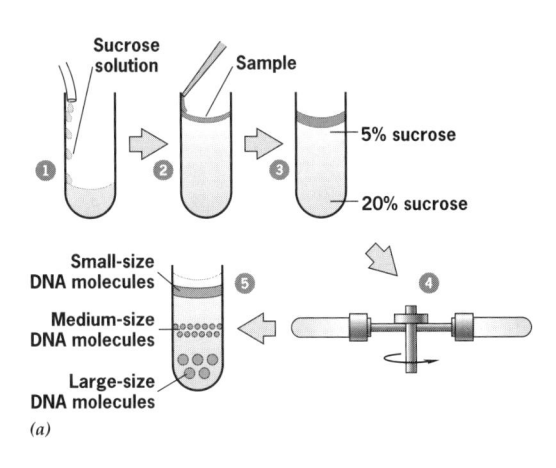

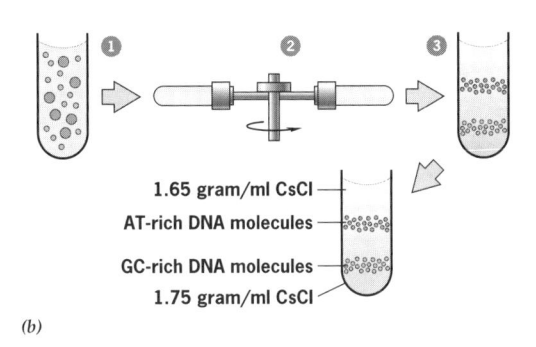

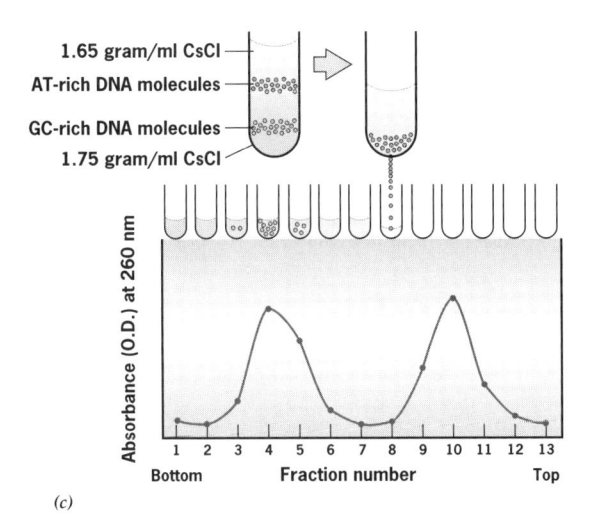

Figure 18.34 Techniques of nucleic acid sedimentation.

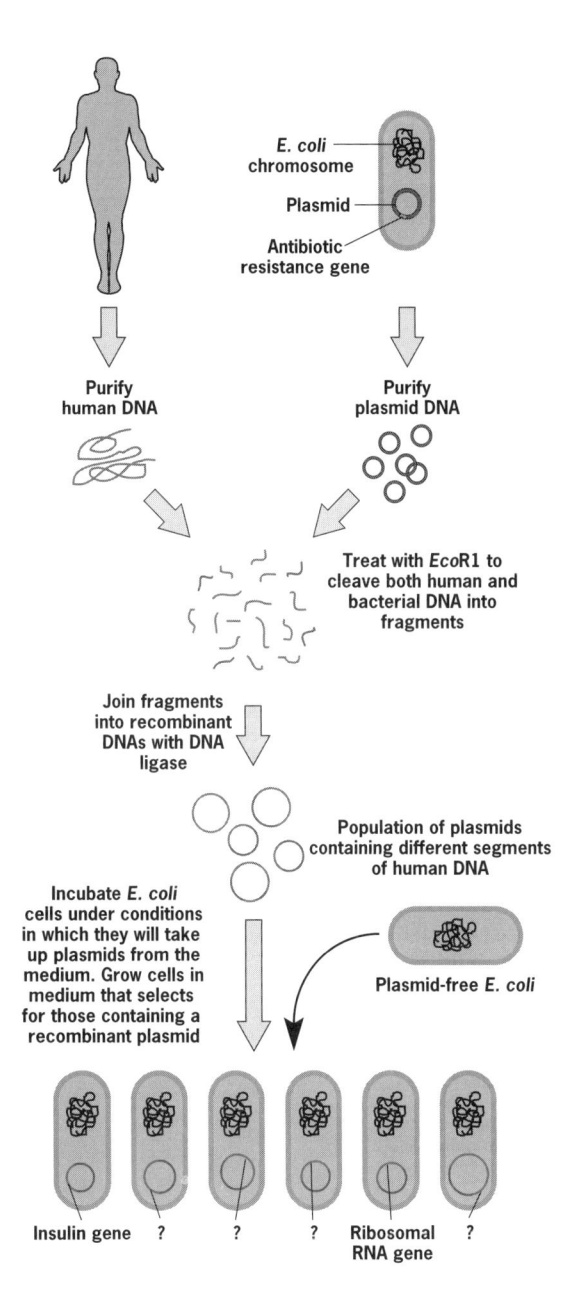

Figure 18.37 An example of DNA cloning using bacterial plasmids.

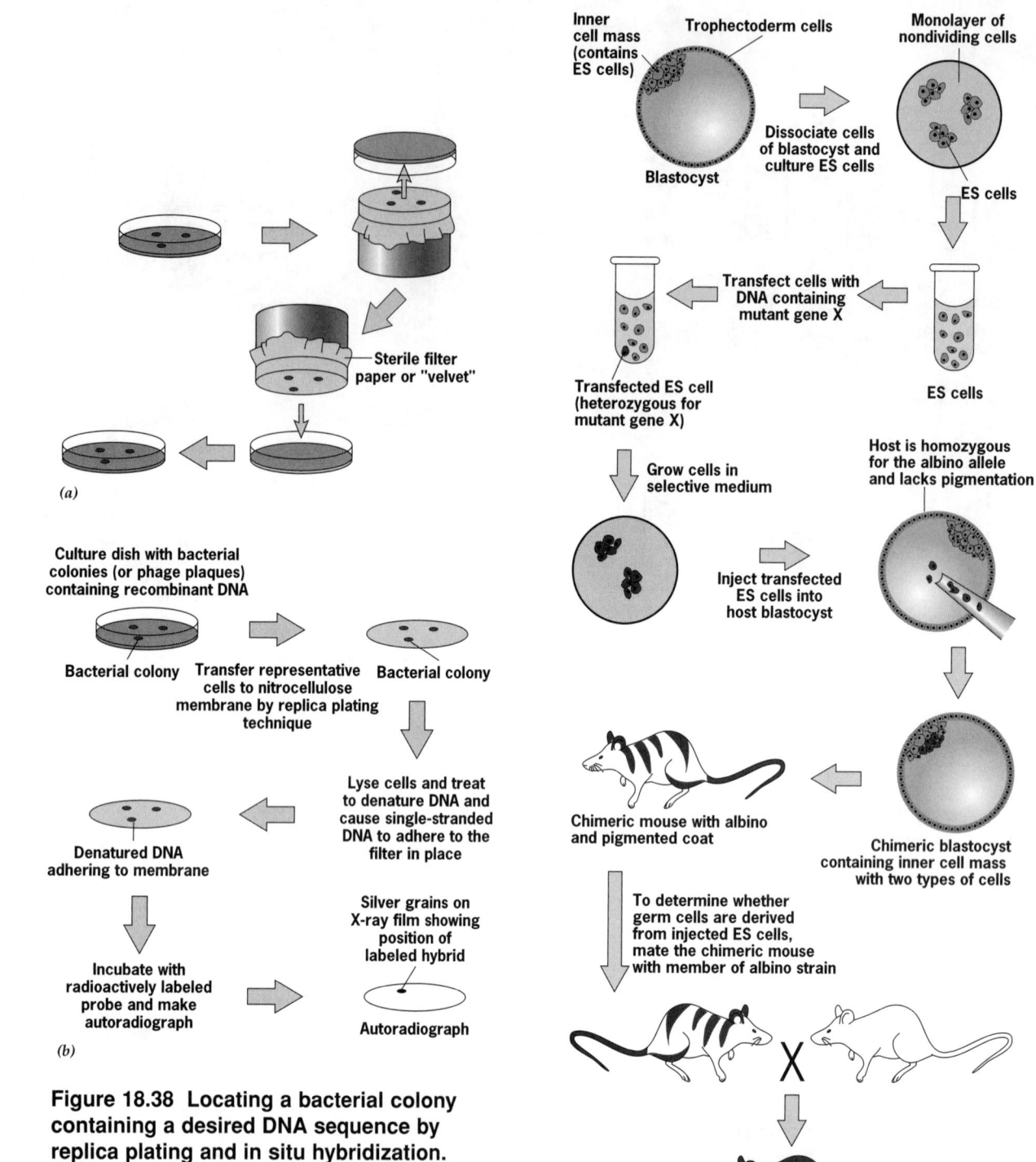

Figure 18.38 Locating a bacterial colony containing a desired DNA sequence by replica plating and in situ hybridization.

Figure 18.45 Formation of knockout mice.

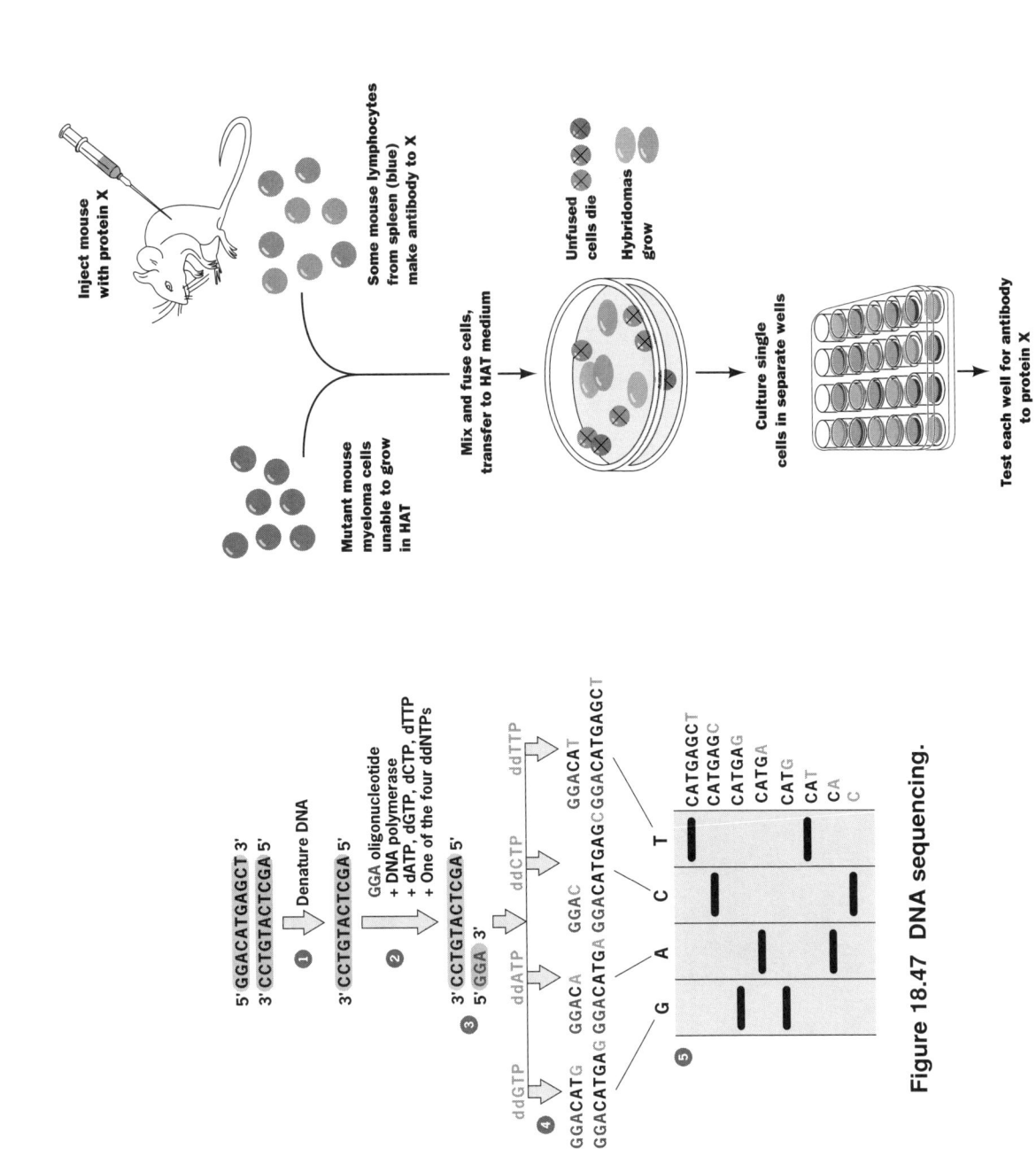

Figure 18.48 The procedure used in the formation of monoclonal antibodies.

Figure 18.47 DNA sequencing.

Figures 18.47 & 18.48